ENCYCLOPEDIA OF EARTH SCIENCE, GEOLOGY AND ENVIRONMENTAL SCIENCE

About the Book

This Book focuses on Earth Science, Geology, Environmental Science and technology for improving environmental quality in every aspect. Environmental scientists work on subjects like the understanding of earth processes, evaluating alternative energy systems, pollution control and mitigation, natural resource management, and the effects of global climate change. Environmental issues almost always include an interaction of physical, chemical, and biological processes.

About the Author

***Dr. Bhawna Garg,** has vast experience of 20+ years in teaching in various government and private organisations, along with throughout first class academic record with Gold Medal in MSc. Statistics,having excellent computer proficiency in the related areas of the various subjects of Statistics like SPSS. Giving guidance to the M.Phil and PHd students of various universities. Taking the responsibilities of publication of Pranjana international journal of management awareness and research development cell.She has authored 9 Books in the field of Statistics, Entrepreneurship, HR and Neural Networks etc. and 30 Research Papers have been published in reputed research journals National as well as International. Several books and research papers have been published abroad. Life member of "Society for reliability engineering, quality and operations management (SREQOM)"Delhi. Member of editorial board of different national and international journals.Phd guide to guide students for Phd JJTU University Rajasthan and Asia Pacific university Udaipur.*

ENCYCLOPEDIA OF EARTH SCIENCE, GEOLOGY AND ENVIRONMENTAL SCIENCE

DR. BHAWNA GARG

AMIGA PRESS INC.

DELHI (INDIA)

AMIGA PRESS INC.
H.O.: 28-A, Indra Park, Chander Nagar
Delhi-110051 (India)
Tel.: 011-22504028, +91-98736-57000
E-mail: amigapress@gmail.com
amigapub@gmail.com
Website : www.amigapress.co.in

This edition is published by Amiga Press Inc.

ISBN: 978-93-84533-46-5

First Published 2019

Printed and bound in India by Replika Press Pvt. Ltd.

Preface

Environmental science is a multidisciplinary academic field that integrates physical, biological and information sciences to the study of the environment, and the solution of environmental problems. Environmental science provides an integrated, quantitative, and interdisciplinary approach to the study of environmental systems. Related areas of study include environmental studies and environmental engineering. Environmental studies incorporates more of the social sciences for understanding human relationships, perceptions and policies towards the environment.

Environmental engineering focuses on design and technology for improving environmental quality in every aspect. Environmental scientists work on subjects like the understanding of earth processes, evaluating alternative energy systems, pollution control and mitigation, natural resource management, and the effects of global climate change. Environmental issues almost always include an interaction of physical, chemical, and biological processes.

Environmental Scientists bring a systems approach to the analysis of environmental problems. Key elements of an effective environmental scientist include the ability to relate space, and time relationships as well as quantitative analysis.

Environmental Science came alive as a substantive, active field of scientific investigation in the 1960s and 1970s driven by the need for a multi-disciplinary approach to analyze complex environmental problems, the arrival of substantive environmental laws requiring specific environmental protocols of investigation and the growing public awareness of a need for action in addressing environmental problems.

Events that spurred this development included the publication of Rachael Carson's landmark environmental book Silent Spring along with major environmental issues becoming very public, such as the 1969 Santa Barbara oil spill, and the Cuyahoga River of Cleveland, Ohio, "catching fire" and helped increase the visibility of environmental issues and create this new field of study. This encyclopedia will prove to be useful and informative for students, research scientists, engineers and teachers in the field of environment science and technology.

–Author

Contents

1

Effect of Environment

ATMOSPHERE

The atmospheric chemistry studies the chemical composition of the natural atmosphere, the way gases, liquids, and solids in the atmosphere interact with each other and with the earth's surface.

For convenience of study, atmospheric scientists divide the atmosphere as if it consists of 4 layers. The division is mainly due to temperature variations as the altitude increases.

The 4 layers according to the variation of temperature are:

1. Ionosphere
2. Mesosphere
3. Stratosphere
4. Troposphere

In the given figure as shown below, 100 km is the thermosphere and ionosphere where the temperature increases from 200 K at 100 km to 500 K at 300 km. The temperature goes even higher as the altitude increases. activity as the altitude decrease. In the outer space, most particles consist of single atoms, H, He, and O etc. At lower altitude (200–100 km), diatomic molecules N_2, O_2, NO etc are present. The ionosphere is full of electrically charged ions. The UV rays ionizes these gases. The major reactions are

In the ionosphere: $O + h\upsilon \rightarrow O^+ + e$

$$N + h\upsilon \rightarrow N^+ + e$$

In the neutral thermosphere:

$$N + O_2 \rightarrow NO + O$$
$$N + NO \rightarrow N_2 + O$$
$$O + O \rightarrow O_2$$

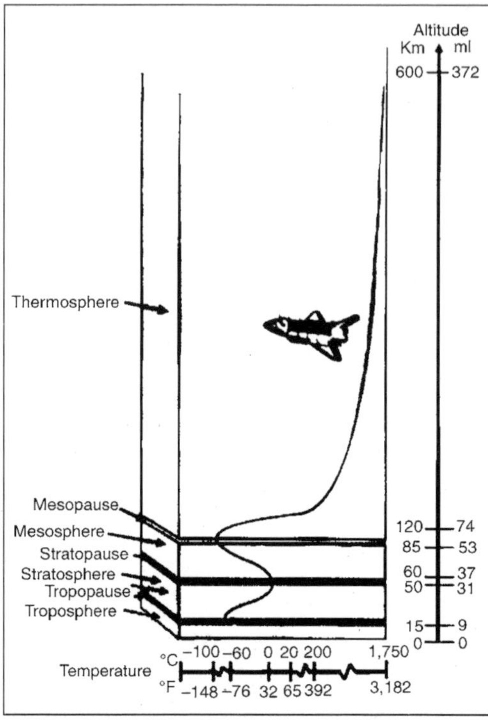

Beyond the neutral thermosphere is the ionosphere and exosphere. These layers are of course interesting for space explorations and environmental concerns and space sciences. The atmosphere in the outer space is more like a plasma than a gas. Below the thermosphere is the mesosphere (100–50 km) in which the temperature decreases as the altitude increase. In this region, OH, H, NO, HO_2, O_2, and O_3 are common, and the most prominent chemical reactions are:

$$H_2O + h\upsilon \rightarrow OH + H$$

$$H_2O_2 + O \rightarrow OH + OH.$$

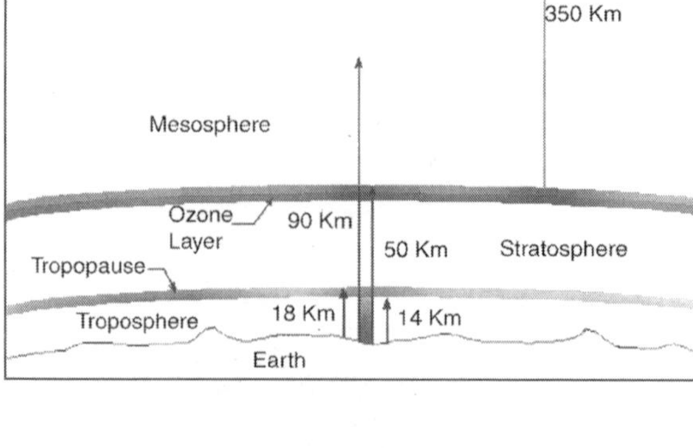

Below the mesosphere is the stratosphere, in which the temperature increases as the altitude increase from 10 km to 50 km. In this region, the following reactions are common:

$$NO_2 \rightarrow NO + O$$
$$N_2O \rightarrow N_2 + O$$
$$H_2 + O \rightarrow OH + H$$
$$CH_4 + O \rightarrow OH + CH_3$$

Air flow is horizontal in the stratoshpere. A thin ozone layer in the upper stratosphere has a high concentration of ozone. This layer is primarily responsible for absorbing the ultraviolet radiation from the sun. The ozone is generated by these reactions:

$$O_2 + h\,v \rightarrow O + O$$
$$O_2 + O \rightarrow O_3$$

The troposphere is where all weather takes place; it is the region of rising and falling packets of air. The air pressure at the top of the troposphere is only 10% of that at sea level (0.1 atmospheres). There is a thin buffer zone between the troposphere and the next layer called the tropopause.

The major components in the region close to the surface of the Earth are N_2 (78%), O_2 (21%), Ar (1%) with variable amounts of H_2O, CO_2, CH_4, NO_2, NO_2, CO, N_2O, and O_3. The ozone concentration in this layer is low, about 8% of the total ozone in the atmosphere is in the troposphere.

All the weather phenomena happen in the troposphere. If you want to know the current weather con dition, check the AccuWeather network for a good view of the troposphere.

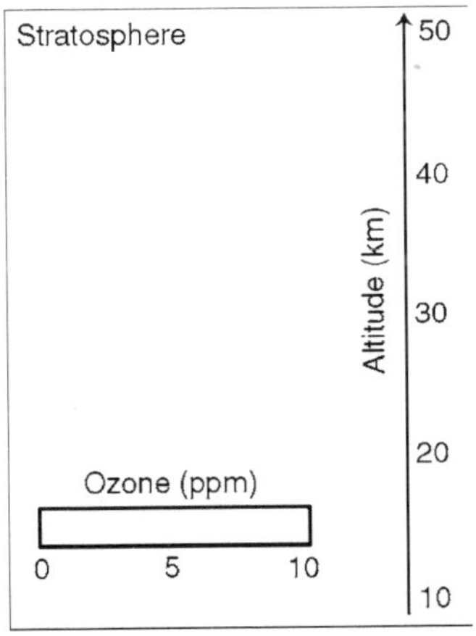

GASES IN ATMOSPHERE

From the atmospheric science viewpoint, interactions of all gasses amoung themselves and their interaction with the environmental elements are of interest. However, for identification purposes, we need to identify the gases produced by man-made process (industry).

Some of the gases due to human activities are:

- Carbon dioxide result from the excess burning of carbon-containing fuel.
- Carbon monoxide produced by automobiles. This orderless and Colourless gas is very toxic.
- Ozone produced in the exhaust of internal combustion engine, and the variation of ozone concentration in the stratosphere.
- Nitrogen oxides such as NO, NO_2, N_2O_4; due to the production of NO in the internal combustion engin.
- Methane gas produced due to treatments of large amount of waste. Sulfur oxides produced in mining operation and in the combustion of sulfur containing fuel. Sulfur oxide causes the so called acid rain problem.
- Chlorofluorocarbons (CFC) are gases used as refrigurant. When disposed into the atmosphere, they cause the ozone concentration to decrease.

Water vapour is also considered a greenhouse gas, but it is also generated by nature contineously due to radiation from the Sun. Of course, when water vapour condense into a liquid, much energy is released in the exthermal process. Condensation of watervapour causes storms and many of the weather phenomena.

STRATOSPHERE AND TROPOSPHERE

One of the most important developments in the past several years has been the realization that the next phase in atmospheric chemistry, dynamics, radiation, and climate research will demand a far more integrated application of modeling, satellite retrievals, and high-resolution, in situ observations. This realization stems from two different causes:

The first is the increasing awareness that scientific research must develop forecasts that are tested and trusted internationally in such areas as climate change, UV dosage levels resulting from ozone changes in the stratosphere, toxicity levels stemming from the urban-to-regional emission of nitrates, sulfates, organics, soot and heavy metals, etc. This demands an integration of transport and photochemical mechanisms that go well beyond the normal alliance that stems from curiosity-driven studies that operate without regard for timetables or societal demands.

The second is the increasing level of sophistication in modeling, data assimilation, and technology related to observational systems, whether they be remote or in situ. This enables the scientific community to investigate in finer granularity the cause and effect relationships in the study of atmospheric chemistry and dynamics. In response to this new realization, the Anderson Group has pursued a course of action based upon a set of objectives of critical importance in today's world.

TROPOSPHERE AND STRATOSPHERE

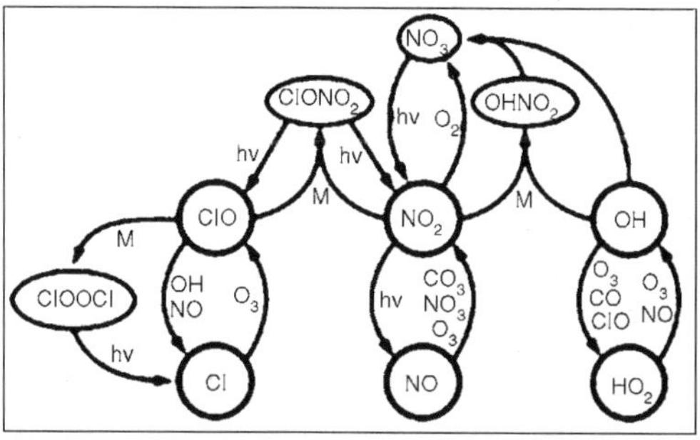

Hypothesis

The boundary condition for water vapour entering the stratosphere in the tropics (irreversible passage through the 390 K è surface) is set by the vertical/horizontal temperature structure of the cold trap region over the Western Tropical Pacific through which the quasi-horizontal passage of air provides adequate residence time (coupled potentially with multiple traversals of these cold pools) to desiccate the air to a vapour pressure corresponding to the local, seasonally dependent, temperature minimum.

Understanding the processes governing stratospheric water vapour is broadly recognized as a dominating scientific need. Changes in stratospheric water vapour under future climate regimes would strongly affect both radiative balance and ozone chemistry.

In particular, increases in stratospheric water content could dramatically increase ozone losses by:

- Increasing the cold aerosol surface area that draws down the NO_x/NO_y ratio thereby increasing the fraction of reactive halogens and hydrogen species portioned into free radical form (ClO, BrO, IO, OH and HO_2)
- Promoting formation of the polar stratospheric clouds on which catalytic ozone loss chemistry occurs
- Increasing the source of HO_x radicals
- Radiative cooling of the lower stratosphere

Because of a lack of measurements, however, debate continues about a wide array of potential scenarios by which air entering the stratosphere is dehydrated. As an illustrative example, we note that while the average saturation mixing ratio of the tropical tropopause in January is 3 ppmv, the average water vapour content of air entering the stratosphere is 4 ppmv. Yet, nearly one quarter of the inner tropics in January has a cold point tropopause temperature corresponding to 2 ppmv (or less) co-located with the region of the most intense convection; so what fraction of the air entering the stratosphere (as it moves zonally and quasi-horizontally with convection superimposed on this background motion) has passed through this cold region?

OZONE AND UV DOSAGE

Societal Objectives

The mission statement identifies as a first order societal issue: "To understand and protect our home planet." A central challenge facing the United States and other countries in the twenty-first century will be to enhance human well-being in a world where growing populations and the drive to improve living standards place potentially huge demands on natural resources and the environment. Whether we succeed or fail in meeting this challenge will be determined, in part, by how we respond to immediate demands to address human health and economic growth in the context of the wide range of crucial, environmentally related decisions made every day by insurance companies, water resource managers, agribusiness, households, city planners, public health officials, and countless others. Rising to this challenge will entail using natural resources as efficiently as possible, devising practical solutions that meet our immediate needs and also provide for long-run economic growth, while maintaining the environmental systems on which life depends.

As we have seen, the Earth's system can be fragile. Stratospheric ozone loss has had profound implications for human health. It has lead to disfiguring and deadly skin cancers that occur as a result of increased ultra-violet radiation reaching the earth's surface. Losses over northern mid-latitudes will affect the population Centres of Europe and North America.

Isolating the cause is complicated by the fact that the photochemical lifetime of ozone in this region is comparable to the dynamical residence time and the dynamical residence time is highly variable depending on season, altitude, longitude, and latitude. Careful measurements and analyses are required to identify and quantify the role of both chemistry and dynamics in determining mid-latitude ozone conce-ntrations.

To guide wise public policy decisions that continue to improve human and economic conditions and to clarify public debate, it is necessary to restructure the science and engineering framework addressing the biological, chemical, and physical integrity of our surroundings. Private-sector and governmental decisions will be made regarding air, water, and living systems that will fundamentally affect our nation's health and its economic and environmental vitality.

AIRBORNE MISSIONS

Airborne Platform

The Anderson Group instruments constitute an important component of the observational array required to address both the AURA calibration/validation requirements and the hypotheses that constitute the scientific objectives of the Programmes. Consideration of AURA colLabourative science and the scientific objectives of TC^4 serves to identify required characteristics of the platform used in the airborne field operation that constitutes the core of this effort:

- Ability to carry the proper complement of instruments
- Ability to contiguously cover the altitudes required from the boundary layer to the lower stratosphere
- Ability to operate in convectively active regions and to engage in sophisticated storm avoidance
- Ability to minimize operational complexity for foreign deployments
- Ability to accept new instruments and to accept multiple instruments for intercomparison
- Ability to combine *in situ* and remote observations to increase the colLabourative overlap between aircraft and satellites.

LASER DEVELOPMENT

A problem of fundamental importance to the NASA vision and mission to improve life here and to understand and protect our home planet, involves a new strategic plan for establishing the quantitative link between rapidly increasing urban/regional emission of aerosol precursors (nitrate, sulfate, heavy metal, organic and elemental carbon) and changes in the optical properties of the atmosphere that constitute a powerful climate forcing.

Dissecting the mechanisms underlying these changes requires measurements that span temporal scales from seconds to years and spatial scales from tens of meters to thousands of Kilometres, to provide an effective union with A-Train satellites.

This research will advance an array of innovative new developments in cavity enhanced spectroscopic techniques employing new miniaturized laser systems coupled with high finesse cavities that will provide a new generation of observations from NASA UAVs.

Central to these efforts will be the advancement of:

- Difference frequency generation (DFG), both singly resonant and intracavity,
- Pulsed fibre lasers using emerging master oscillator power amplifier (MOPA) configurations, micro-etalons, and double-clad fibres, and
- Tailored quantum cascade lasers using newly developed beam epitaxy approaches.

In conjunction with these laser developments, we will advance developments in flight electronics with emphasis on scalable distributed processing and data acquisition necessary for the inclusion of a scientifically potent instrument array within the weight and volume restrictions of UAVs. A key component of this research endeavor is the objective of unifying the sub-orbital and satellite components of the Programmes.

COMBUSTION RELATED AIR POLLUTION

AIR POLLUTION

Air pollution, both indoor and outdoor, is a significant cause of health problems worldwide. Urban and rural outdoor environments contain infections, allergens, irritants and chemical toxins that can reduce the quality of life and cause disease. Inhaled air pollution is directed at the the nose, throat and lungs. The exposed airway allows hazardous pollutants to enter the body and all tissues are ultimately exposed.

The tolerance for environmental destruction is ancient and human history is littered with civilizations that failed because humans indiscriminately exploited natural resources and spoiled their own nest. Humans adapt easily to deteriorating conditions and will persist in following daily routines even when air pollution is severe, traffic is congested, water and food supplies are at risk, and social order is unstable.

Fine-particulate air pollution is known to contribute to cardiovascular and lung disease, increasing the risk of heart attack and a heart-related death. Researchers at Brigham Young University and Harvard School of Public Health compared changes in air pollution from 1980 to 2000 with residents' life expectancies, They concluded that a reduction in air pollutants in 51 U.S. cities between 1980 and 2000 added an average of five months to life expectancy.

Residents in cities that made the most significant improvements in air quality, such as Pittsburgh, PA, lived almost 10 months longer. For every microgram per cubic meter decrease in fine-particulate air pollution, life expectancies rose by more than seven months. Airborne chemicals contaminate food and water. They may be ingested and are also collected in the nose and throat and swallowed, often in mucus that attempts to protect exposed surfaces.

Airborne chemicals entering the digestive system include well-known toxins such as pesticides, organophosphate, PCBs, dioxin, arsenic, cadmium, lead, and mercury. In addition occupational exposures to airborne pathogens can be intense and can cause cancer.

While ambient air pollution is a major concern, indoor air can be more polluted than outdoor air. Building materials and furnishings are a source of volatile chemicals. A decrease in indoor air quality is the result of reduced ventilation and efficient construction practices, sealing homes, stores and office buildings from the outdoor environment.

Smoking tobacco remains a personal method of producing air pollution that remains popular worldwide, despite overwhelming evidence that tobacco smoke produces a long list of disabling and fatal diseases. It is estimated that 30% of all fatal cancers could be prevented if tobacco smoking were eliminated from the list of air pollutants.

The really sad part of our current predicament is that all the right ideas for creating a healthy environment have been around for decades and have been clearly articulated in many forms by a host of intelligent people. The right ideas involve unselfish and compassionate Behaviour. The right ideas involve long-term planning, conservation and deep commitment to preserving the natural world. Without a healthy natural environment, there will be few or no healthy humans.

Our big environmental problems are built from many small, personal decisions - little mistakes that add up over time. If there is a solution, it will emerge from the collective value of millions of better decisions made by individuals all over the globe. The environmental action plan is to think globally and act locally - it does make sense.

AIR POLLUTANTS

In cities across the globe, the personal automobile is the single greatest polluter, as emissions from a billion vehicles on the road add up to a planet-wide problem. Driving a private car is a typical citizen's most air polluting activity. The negative effects of automotive emissions are maximum when you sit in traffic surrounded by cars, their engines idling. Everyone in a traffic jam is getting poisoned.

The Combustion Process Gasoline and diesel fuels are mixtures of hydrocarbons (made of hydrogen, oxygen and carbon atoms.) Hydrocarbons are burned by combining with oxygen. Nitrogen and sulphur atoms are also present and combine with oxygen when burned to produce gases. Automotive engines emit several types of pollutants.

Typical Engine Combustion

Fuel + Air ⇒ Hydrocarbons + Nitrogen Oxides + Carbon Dioxide + Carbon Monoxide + water Hydrocarbon emissions are fragments of fuel molecules, only partially burned. See Toxicity of Benzene and other Hydrocarbons in exhaust.

Hydrocarbons react in the presence of nitrogen oxides and sunlight to form ground-level ozone, a major component of smog. Ozone irritates the eyes, nose, throat and damages the lungs. A number of exhaust hydrocarbons are also toxic, some with the potential to cause cancer.

Nitrogen Oxides

Under high pressure and temperature conditions in an engine, nitrogen and oxygen atoms react to form nitrogen oxides. Nitrogen oxides, like hydrocarbons, are precursors to the formation of ozone and contribute to acid rain. Catalytic converters in car exhaust systems break down heavier nitrogen gases, forming nitrous oxide (NO_2) - 300 times more potent than carbon dioxide as a greenhouse gas. Nitrous oxide makes up about 7.2 per cent of the gases that cause global warming. Vehicles with catalytic converters produced nearly half of that nitrous oxide. Nitrous oxide also comes from nitrogen-based fertilizers and manure from farm animals.

Carbon Monoxide

Carbon monoxide (CO) is a Colourless, odourless, poisonous gas, a product of incomplete burning of hydrocarbon-based fuels. Carbon monoxide consists of a single carbon atom and a single oxygen atom linked together (CO), the product Carbon monoxide of incomplete combustion of fuel. Most CO is produced when air-to-fuel ratios are too low in the engine during vehicle starting, when cars are not tuned properly, and at higher altitudes, where thin air reduces the amount of oxygen available for combustion. Two-thirds of the carbon monoxide emissions come from transportation sources, with the largest contribution coming from cars. In urban areas, the passenger vehicle contribution to carbon monoxide pollution can exceed 90%.

Carbon Dioxide

Environmental Protection Agency (EPA) originally viewed carbon dioxide as a product of "perfect" combustion, but now views CO_2 as a pollution concern. Carbon dioxide is a greenhouse gas that traps the earth's heat and contributes to Global Warming.

Evaporative Emissions

Hydrocarbon pollutants also escape into the air through fuel evaporation - evaporation causes significant hydrocarbon pollution from cars on hot days when ozone levels are highest. Evaporative emissions occur several ways:

Diurnal

Gasoline evaporation increases as the temperature rises during the day, heating the fuel tank and venting gasoline vapors.

Running Loses

The hot engine and exhaust system can vaporize gasoline when the car is running.

Sitting Evaporation

The engine remains hot for a period of time after the car is turned off, and gasoline evaporation continues when the car is parked.

Adding Fuel

Gasoline vapors are always present in fuel tanks. These vapors are forced out when the tank is filled with liquid fuel.

Benzene is the main toxin in the hydrocarbon fraction of exhaust. Benzene and other less known hydrocarbons are produced in petroleum refining, and are widely used as solvents and as materials in the production of various industrial products and pesticides. Benzene also is found in gasoline and in cigarette smoke. Other environmental sources of benzene include gasoline (filling) stations, underground storage tanks that leak, wastewater from industries that use benzene, chemical spills, and groundwater next to landfills containing benzene. Exposure to benzene is related to the development of leukemia and lymphoma. Benzene has a suppressive effect on bone marrow and it impairs blood cell maturation and amplification.

Polycyclic Aromatic Hydrocarbon (PAH)

PAHs are a group of chemicals that are formed during the incomplete burning of coal, oil and gas, garbage, or other organic substances. PAHs can be man-made or occur naturally. A few of the PAHs are used in medicines and to make dyes, plastics, and pesticides. They are found throughout the environment in the air, water and soil. There are more than 100 different PAH compounds. Although the health effects of the individual PAHs vary, the following 15 PAHs are considered as a group with similar toxicity: acenaphthene, acenaphthylene, anthracene, benzanthracene, benzopyrene, benzofluoranthene, benzoperylene, benzofluoranthene, chrysene dibenzanthracene, fluoranthene, fluorene, indenopyrene, phenanthrene, pyrene.

Long term solutions require reduced combustion of all kinds. While vehicles with new energy sources such ethanol, biofuels, propane and natural gas can contribute to reduced air pollution, their benefit is limited if vehicle use continues at current intensities. If you pay more money to buy a hybrid car, but drive it more, you have contributed little to solving air pollution problems. If you buy a gas guzzling clunker and use only one gallon of gas to 20 miles each week, you have contributed more to the solution.

The problem with all alternative fuels is that the manufacture of fuels requires energy, distribution with a manufacturing infrastructure that consume energy, often derived from burning fossil fuels. No alternative fuel is ideal. While ethanol has been championed as an alternative to petroleum fuels, it mainly helps to reduce dependency on oil producing countries. When ethanol is made from corn, more than 75% of its energy value must be spent on its production. Burning ethanol still produces carbon dioxide. Climate change with extreme weather events threatens corn production in the US. The new competition between hastily constructed ethanol plants and food production became an international issue in 2008. Other approaches to new fuels made from non-food plant sources are part of the solution.

Hydrogen

Ultimately cars might burn hydrogen in fuel cells, but despite working prototypes, a hydrogen economy is a distant fantasy. There are many problems to be solved before hydrogen can replace fossil fuels as a portable energy source. The biggest problem is that producing hydrogen requires a large amount of energy. In Canada, there are opportunities to dam rivers and produce electricity with falling water, a non polluting, renewable energy resource. A more problematic energy source would be be nuclear reactors that "burn" uranium or plutonium.

Even if new non-polluting energy sources are developed, hydrogen storage and distribution requires a new, very expensive infrastructure that could replace gasoline and diesel fuels. With once rich countries such as the USA on the verge of bankruptcy and facing the extensive repairs of already aging, derelict infrastructures, adding a new, unprecedented development costs seems unlikely. Unless, of course the priorities in the

US shift dramatically. The US could adopt a sane, smart strategy, reduce its military budget by 60% and invest the money and skills in rebuilding the country's infrastructure with new sustainable energy sources.

GLOBAL ENVIRONMENTAL PROBLEMS

INTRODUCTION

Ethics is understood to be the domain of enquiry that explores what is right or wrong, obligatory or non-obligatory, or when responsibility attaches to human Behaviour. Are there features of global environmental problems that call for classifying them as essentially ethical problems with even greater force than some local or regional environmental problems? If so, what are these features?

Why is this important? If some global environmental problems are essentially ethical problems, then some of the excuses that nations often use to justify not reducing their contributions to the global environmental problems are ethically problematic. This is so because ethical obligations entail duties and responsibilities to others with the result that national policies may not be justified on national interests alone. That is, if there are ethical obligations to others to cease causing harm, then policy options must be responsive to these obligations.

The features of global environmental problems that strongly call for classification as ethical problems. Some global environmental problems are caused by people in one part of the world but are most harshly experienced by others who are separated from those causing the problem by great space and time.

In addition, those most vulnerable to global environ-mental problems are often least responsible for causing them. Climate change is a strong example of this feature of global environmental problems. Those most vulnerable to climate include many of the poorest people in the world who are among those least responsible for emitting the greenhouse gases that are causing the problem. The same separation between victim and cause is also true of other global environmental problems, including upper atmospheric ozone depletion, some ocean degradation problems including destruction of global fisheries, some biodiversity and forest problems, and exposure to toxic substances that are deposited around the world by long-range air transport.

In the case of global fisheries depletion, for instance, fishing trawlers from some countries often go great distances to harvest fish for domestic consumption, yet the harshest consequences of depleted global fisheries are often experienced by subsistence fisherman in poor countries far away from the home port of the fishing boats.

This separation of space and time and between cause and impact of course can also happen in more local environmental problems such as water pollution, but the distances between those causing the problem and those who suffer the consequences are most problematic at the global scale because of the magnitude of the distances at this scale.. With the rise in truly global environmental problems in the last few decades, it is now abundantly clear that Behaviour of people in one part of the world can gravely affect the human health, economic viability and overall welfare of others far away, but their suffering is almost invisible to those who cause the problem.

Several implications for ethical reasoning follow directly from this separation of distance and time between cause and effect. First, our intuitive sense of ethical responsibility for harms that we cause others is harder to trigger when those who will be harmed by our action are people we will not ever know and are from a different culture. Yet ethics and morality require us to refrain from harming all others who may suffer because of our Behaviour even if they are separated from us in time and space.

Secondly, because the actual harm that we are causing to others may be in a particular part of the world that is uniquely vulnerable to global environmental change but is unknown to us, we not only don't know the people we are harming but have no idea where the locus of the worst harm is located. For instance, particularly vulnerable villages to climate change impacts may turn on the geographical features where the village is located such as dependence on glacier water flows in the summer coupled with vulnerability to monsoon irregularities. Yet few of us know where these places are. And so in the case of global environmental problems we may be causing great harm to others who we not only do not know personally, but about whom we neither have an idea of their location nor why they are particularly vulnerable to actions we are taking. For this reason, global scale environmental problems create a huge challenge to trigger our sense of ethical responsibility, yet ethics requires us to not harm others through our actions none the less.

A related reason why the global environmental problems demand classification as ethical problems stems from the fact that some actions that cause some environmental damage are often justified on the basis of cost- benefit analysis. Yet this justification is ethically problematic when those who benefit from an action that causes harm have no relation with those who are harmed. For instance, government may tolerate inevitable pollution from a land fill by identifying the benefits that citizens receive by having a place to take their garbage.

But in the case of global environmental problems, those who are harmed by actions do not enjoy the benefits that come from taking the action. A harms- benefits justification for environmental degradation, if ethically justifiable at all, only works where those who will suffer the harms and those who benefit from the action that causes the harm are in the same community of interests. In the case of global environmental problems, those who are harmed are rarely made better by the benefits of the action that cause the harm.

The second reason why global environmental problems call for classification as ethical problems stems from the fact that their consequences are often catastrophic to those who are most harmed by them. Climate change, for instance, directly threatens human life and health and resources to sustain life as well as species of plants and animals and ecosystems around the world.

The harms include deaths from disease, droughts, floods, heat and intense storms and damage to homes and villages from rising oceans and intense storms, adverse impacts on agriculture, social disputes caused by diminishing natural resources, sickness from a variety of diseases, the inability to rely upon traditional sources of food, the destruction of water supplies, and the inability to live where one has lived to sustain life. In addition, the very existence of some small island nations is threatened by climate change caused seal level rise. Clearly these impacts are catastrophic for some.

Loss of global fisheries can be devastating to substance fishermen, global desertification and drought are likely to cause mass starvation, loss of atmospheric ozone can cause skin cancer in millions, and long-range transport of persistent organic chemicals can cause deadly diseases and birth defects. Because ethics requires people to be particularly careful to not to harm others when the harm is great, global environmental problems need to be seen as ethical issues.

Governments are expected to protect their citizens from life-threatening dangers. At the local, regional or national scale, citizens can petition their governments to protect them from environmental harms. But at the global level, no government exists whose jurisdiction matches the scale of global environmental problems.

And so, although existing national, regional and local governments have jurisdiction over activities within their boundaries coupled to responsibilities to their citizens, they have no responsibility to those outside their boarders in the absence of binding international law. For this reason, ethical appeals to get governments to require that their citizens not harm non-citizens in other nations are necessary to prompt action.

In fact, according to a leading foreign policy theory called "realism," national governments should limit the scope of their international engagements to the pursuit of national interest alone. According to some "realist" proponents, governments should refrain from acting at the international level for ethical reasons that do not coincide with national interests. Yet global environmental problems require people in one part of the world to consider the interests of others outside the jurisdiction of their national governments. Because there is no duty of national governments to protect foreigners from global environmental problems unless they bind themselves voluntarily in treaties, an appeal to ethical and moral responsibility is particularly important to motivate national actions in regard to global scale environmental problems. Moreover, if global environmental problems actually create ethical obligations, they have ethical duties to mitigate global environmental problems

CFCS IN ATMOSPHERE

There has been a movement since the late 1970s to ban CFCs because of their destructive effect on the ozone layer. This damage was discovered by Sherry Rowland and Mario Molina, who first published a paper suggesting the connection in 1974. It turns out that one of CFCs' most attractive features- their unreactivity-has been instrumental in making them one of the most significant pollutants. CFCs' lack of reactivity gives them a lifespan which can exceed 100 years in some cases. This gives them time to diffuse into the upper stratosphere. Here, the sun's ultraviolet radiation is strong enough to break off the chlorine atom, which on its own is a highly reactive free radical. This catalyses the break up of ozone into oxygen:

$$Cl + O_3 \rightarrow ClO + O_2$$

$$ClO + O \rightarrow Cl + O_2$$

CFCs are a problem because the chlorine is regenerated at the end of these reactions, making it able to repeat this reaction millions of times. This disrupts the ionic structure of oxygen atoms, causing their lowest ionisation energy to drop. This reaction is believed to be the cause of the ozone hole observed over the poles and upper latitudes of the Earth. Decreases in stratospheric ozone have been linked to increases in skin cancer. In 1975, Oregon enacted the world's first ban of CFCs. By 1985, scientists observed a dramatic seasonal depletion of the ozone layer over Antarctica, which led the United Nations to take action in 1987. International attention to CFCs resulted in a meeting of world diplomats in Montreal in 1987. They forged a treaty, the Montreal Protocol, which called for drastic reductions in the production of CFCs. On March 2, 1989, 12 European Community nations agreed to ban the production of all CFCs by the end of the century. In 1990, diplomats met in London and voted to significantly strengthen the Montreal Protocol by calling for a complete elimination of CFCs by the year 2000. By the year 2010 CFCs should be completely eliminated from developing countries as well

FREON

Freon is a trade name for a group of chlorofluorocarbons used primarily as a refrigerant. The word Freon is a registered trademark belonging to Dupont. Prior to their implication as a major contributor to ozone depletion (and a minor contributor to the greenhouse effect), freons were also used as aerosol propellants and to foam up polymer materials.

Freon was developed as a safer alternative to toxic gases such as, ammonia (NH_3), methyl chloride (CH_3Cl), and sulfur dioxide (SO_2), that were used as refrigerants prior to the discovery of freon. Despite the controversy, R12 works

well and is very safe. Its inventor, Thomas Midgley, actually breathed in a lungful and exhaled over a candle to demonstrate its nontoxicity and nonflammability. Before CFCs were banned by developed nations, a 12 ounce (340 g) can of freon (R12) cost about $1; following the ban the cost of freon rose dramatically. People now go through considerable trouble to recycle freon from existing sources. As of 2002, freon cost about $85 per lb ($187 per kg). Freon is most frequently sold in 12 ounce (340 g) cans and 30 pound (13.6 kg) canisters.

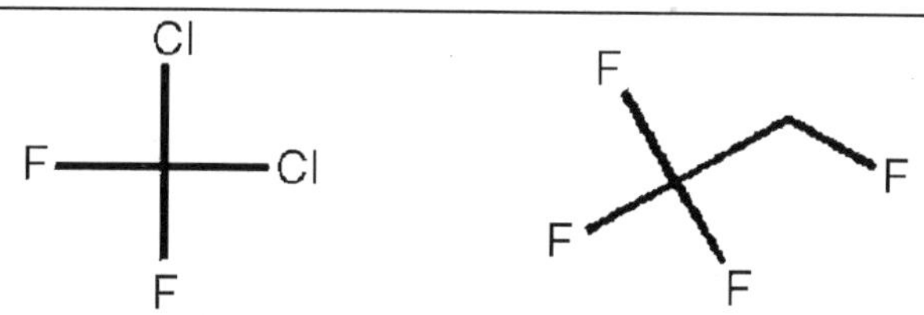

Dichlordifuoromethane "Freon 12" 1,1,1,2-Tetrafluoroethane "Suva 134a"

Freon 12 → (also known as R–12) was a popular re-frigerant before it was banned. The chemical formula is $CCl_2F_2$and its proper chemical name is dichloro-difluoromethane. When people refer to freon without specifying a number, it is likely that they mean freon-12.

Suva 134a → (also known as R–134 or R–134a) is the most common replacement for R–12. Unlike R–12, which is a chlorofluorocarbon, R–134 is a hydro fluorocarbon. The chemical formula is CH_2FCF_3 and the proper chemical name 1,1,1,2-tetrafluoroethane.

	Freon 41	**Fluoromethane**
R-11	CFC-11	Trichlorofluoromethane
R-12	CFC-12	Dichlorodifluoromethane
R-13	CFC-13	Chlorotrifluoromethane
R-22	HCFC-22	Chlorodifluoromethane
R-23	HFC-23	Trifluoromethane

R-113	CFC-113	Trichlorotrifluoroethane
R-114	CFC-114	1,2-dichloro-1,1,2,2-Tetrafluoroethane
R-115	CFC-115	1-chloro-1,1,2,2,2-Pentafluoroethane
R-116	CFC-116	Hexachloroethane
R-134a	HFC-134a	1,1,1,2-tetrafluoroethane
R-227ea	HFC-227ea	1,1,1,2,3,3,3-heptafluoropropane

Freon in copper tubing open to the environment can turn into phosgene gas after coming in contact with extreme heat, such as, while brazing or in a fire situation. Phosgene is a substance that was used as a chemical weapon in World War I. Low exposure can cause irritation, but high levels cause fluid to collect in the lungs, possibly resulting in death.

HALON

Halons are a group of compounds obtained by replacing the hydrogen atoms of a hydrocarbon with halogen atoms, such as bromine or fluorine. Halon 1211 is bromochlo-rodifluoromethane (CF_2BrCl) and Halon 1301 is bromotrifl-uoromethane (CF_3Br). Halons are very stable and are widely used in fire extinguishers where water and other alternatives would be ineffective and dangerous (*e.g.* when dealing with oil fires) or cause unacceptable collateral damage (*e.g.* with electronic equipment.)

Introduction

At high temperatures, halons decompose to release halogen atoms that combine readily with active hydrogen atoms, quenching the flame propagation reaction even when adequate fuel, oxygen and heat remains. The chemical reaction in a flame proceeds as a free radical chain reaction; by sequestering the radicals which propagate the reaction, halons are able to "poison" the fire at much lower concentrations than are required by fire suppressants using the more traditional methods of cooling, oxygen deprivation, or fuel dilution. For example, Halon 1301 flood systems are typically used at concentrations no higher than 6% v/v in air, and can suppress many fires at 2.9% v/v. By contrast, carbon dioxide fire suppression flood systems are operated from 34% (surface-only combustion of liquid fuels) up to 75% (dust traps). Carbon dioxide can cause severe distress at concentrations of 3 to 6%, and has caused death by respiratory paralysis in a few minutes at 10% concentration.

Halon 1301 causes only slight giddiness at its effective concentration of 5%, and even at 15% persons remain conscious but impaired, and suffer no long term effects. (Experimental animals have also been exposed to 2%

concentrations of Halon 1301 for 30 hours per week for 4 months, with no discernible health effects at all.) Halon 1211 also has low toxicity, although it is more toxic than Halon 1301, and thus considered unsuitable for flooding systems.

Fig. 1.1: Halon Canisters Used in a Fire-suppression System

However, halon fire suppression is not completely non-toxic; very high temperature flame, or contact with red-hot metal, can cause decomposition to toxic byproducts. The presence of such byproducts is readily detected because they include hydrobromic acid and hydrofluoric acid, which are intensely irritating. Halons are very effective on Class A (organic solids), B (flammable liquids and gases) and C (electrical) fires, but they are totally unsuitable for Class D (metal) fires, as they will not only produce toxic gas and fail to halt the fire, but in some cases pose a risk of explosion. Halons can be used on Class K (kitchen oils and greases) fires, but offer no advantages over specialised foams.

Nomenclature

Halons are referred to by a system of code numbers similar to (but simpler than) the system used for freons. The first digit specifies the number of carbon atoms in the molecule, the second is the number of fluorine atoms, the third is the chlorine atoms, and the fourth is the number of bromine atoms. If the number includes a fifth digit, the fifth number indicates the number of iodine atoms (though iodine in Halon is rare). (Note that the various atomic species are listed in order of increasing atomic number.) Any bonds not taken up by halogen atoms are then allocated to hydrogen atoms.

For example, consider Halon 1211: C F Cl Br 1 2 1 1

Halon 1211 has one carbon atom, two fluorine atoms, one chlorine atom and one bromine atom. A single carbon only has four bonds, all of which are taken by the halogen atoms, so there is no hydrogen. Thus its formula is CF_2BrCl, and its IUPAC name is therefore bromochlorodi fluoromethane.

Application

Halon 1211 is typically used in hand-held extinguishers, in which a stream of liquid halon is directed at a smaller fire by a user. The stream evaporates under reduced pressure, producing strong local cooling, as well as a high concentration of halon in the immediate vicinity of the fire. In this mode, extinguishment is achieved by cooling and oxygen deprivation at the core of the fire, as well as radical quenching over a larger area. After fire suppression, the halon moves away with the surrounding air, leaving no residue.

Halon 1301 is more usually employed in flooding systems. In these systems, banks of halon cylinders are kept pressurised to about 4 MPa with compressed nitrogen, and ducts lead to the protected enclosure. On triggering, the entire measured contents of one or more cylinders are dumped into the enclosure in a few seconds, through nozzles designed to ensure uniform mixing throughout the room. The quantity dumped is pre-calculated to achieve the desired concentration, typically 3-5% v/v. This level is maintained for some time, typically twenty minutes, to ensure all items have cooled so reignition is unlikely to occur, then the air in the enclosure is purged. During this time the enclosure may be entered by persons wearing SCBA. (There exists a common myth that this is because halon is highly toxic; in fact it is because it can cause giddiness and mildly impaired perception, and also due to the risk of combustion byproducts.) Flooding systems may be manually operated or automatically triggered by a VESDA. In the latter case, a warning siren and strobe lamp will first be activated for a few seconds to warn personnel to evacuate the area. The rapid dumping of halon and consequent rapid cooling fills the air with fog, and is accompanied by a loud, disorienting noise.

Environmental Issues

There is concern that halons are being broken down in the atmosphere to bromine, which reacts with ozone, leading to depletion of the ozone layer (this is similar to the case of chlorofluorocarbons such as freon). These issues are complicated: the kinds of fires that require halon extinguishers to be put out will typically cause more damage to the ozone layer than the halon itself, not to mention human and property damage. However, fire extinguisher systems must be tested regularly, and these tests may lead to damage. As a result, some regulatory measures have been taken, and halons are being phased out in most of the world.

History

Carbon tetrachloride was used in fire extinguishers and glass (anti)-"fire grenades"from the latenineteenth century until around the end of World War Experimentation with halogenated hydrocarbons for fire suppression on military aircraft began at least as early as the 1920s. During World War II, various early halons were in standard use in aircraft by some combatants, but these early halons suffered from excessive toxicity. Nevertheless after the war they slowly became more common in civil aviation as well.

In the 1960s, fluorinated halons became available, and were quickly recognised as one of the most effective fire fighting materials discovered. Much early research with Halon 1301 was conducted under the auspices of the US Armed Forces, while Halon 1211 was initially mainly developed in the UK. By the late 1960s, they were standard in many applications where water and dry powder extinguishers posed a threat of damage to the protected property, including computer rooms, telecommunications switches, Labouratories, museums and art collections. Beginning with warships in the 1970s, halons also progressively came to be associated with rapid knockdown of severe fires in confined spaces with minimal risk to personnel.

By the early 1980s, halons were in common use on aircraft, ships and large vehicles, as well as in computer facilities and galleries. However, concern was beginning to be felt about the possible impact of chlorinated and brominated hydrocarbons on the ozone layer. The Vienna Convention on Ozone Layer Protection did not cover halons, as it was felt that emergency discharge of systems was too small in volume to produce a significant impact, and too important to human safety for restriction. However, by the time of the Montreal Protocol it was realised that discharges during system tests and maintenance accounted for substantially larger volumes than emergency discharges, and so halons were brought into the treaty, but with many exceptions. Halons have been generally phased out and the possession of halon equipment is prohibited in some countries like the Netherlands and Belgium from January 1, 2004, based on the Montreal Protocol and guidelines of the European Union. Production of new stocks has ceased in most (probably all) countries as of 1994. However many countries still require aircraft to be fitted with halon fire suppression systems, as no safe and completely satisfactory alternative has been discovered for this application. There are also a few other highly specialised users.

OZONE DEPLETION

The ozone layer protects the Earth from the ultraviolet rays sent down by the sun. If the ozone layer is depleted by human action, the effects on the planet could be catastrophic.

Ozone is present in the stratosphere. The stratosphere reaches 30 miles above the Earth, and at the very top it contains ozone. The suns rays are absorbed by the ozone in the stratosphere and thus do not reach the Earth.

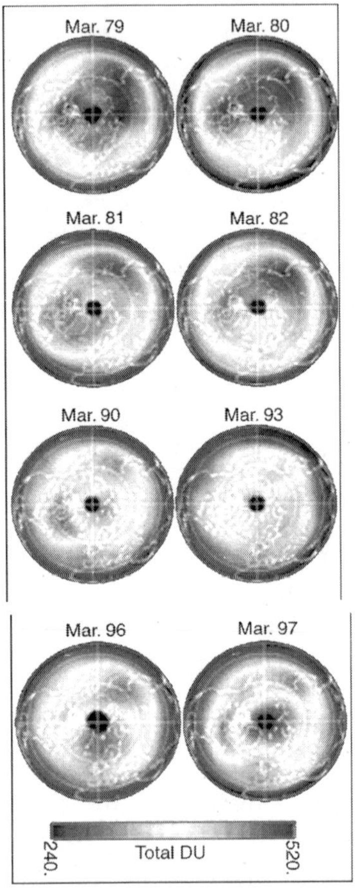

Ozone is a bluish gas that is formed by three atoms of oxygen. The form of oxygen that humans breathe in consists of two oxygen atoms, O_2. When found on the surface of the planet, ozone is considered a dangerous pollutant and is one substance responsible for producing the greenhouse effect.

The highest regions of the stratosphere contain about 90% of all ozone. In recent years, the ozone layer has been the subject of much discussion. And rightly so, because the ozone layer protects both plant and animal life on the planet. The fact that the ozone layer was being depleted was discovered in the mid-1980s. The main cause of this is the release of CFCs, chlorofluorocarbons.

Antarctica was an early victim of ozone destruction. A massive hole in the ozone layer right above Antarctica now threatens not only that continent, but many others that could be the victims of Antarctica's melting icecaps. In the future, the ozone problem will have to be solved so that the protective layer can be conserved.

CAUSES

Only a few factors combine to create the problem of ozone layer depletion. The production and emission of CFCs, chlorofluorocarbons, is by far the leading cause.

Many countries have called for the end of CFC production because only a few produce the chemical. However, those industries that do use CFCs do not want to discontinue usage of this highly valuable industrial chemical.

CFCs are used in industry in a variety of ways and have been amazingly useful in many products. Discovered in the 1930s by American chemist Thomas Midgley, CFCs came to be used in refrigerators, home insulation, plastic foam, and throwaway food containers. Only later did people realise the disaster CFCs caused in the stratosphere. There, the chlorine atom is removed from the CFC and attracts one of the three oxygen atoms in the ozone molecule. The process continues, and a single chlorine atom can destroy over 100,000 molecules of ozone.

In 1974, Sherwood Rowland and Mario Molina followed the path of CFCs. Their research proved that CFCs were entering the atmosphere, and they concluded that 99% of all CFC molecules would end up in the stratosphere.

When the ozone layer hole was discovered over Antarctica, was the proof truly conclusive. At that point, it was hard to question the destructive capabilities of CFCs.

Even if CFCs were banned, problems would remain. There would still be no way to remove the CFCs that are now present in the environment. Clearly though, something must be done to limit this international problem in the future

EFFECTS

Even minor problems of ozone depletion can have major effects. Every time even a small amount of the ozone layer is lost, more ultraviolet light from the sun can reach the Earth.

Every time 1% of the ozone layer is depleted, 2% more UV-B is able to reach the surface of the planet. UV-B increase is one of the most harmful consequences of ozone depletion because it can cause skin cancer. The increased cancer levels caused by exposure to this ultraviolet light could be enormous. The EPA estimates that 60 million Americans born by the year 2075 will get skin cancer because of ozone depletion. About one million of these people will die.In addition to cancer, some research shows that a decreased ozone layer will increase rates of malaria and other infectious diseases. According to the EPA, 17 million more cases of cataracts can also be expected.

The environment will also be negatively affected by ozone depletion. The life cycles of plants will change, disrupting the food chain. Effects on animals will also be severe, and are very difficult to foresee. Oceans will be hit hard as well. The most basic microscopic organisms such as plankton may not be able to survive. If that happened, it would mean that all of the other animals that are above plankton in the food chain would also die out. Other ecosystems such as forests and deserts will also be harmed.

The planet's climate could also be affected by depletion of the ozone layer. Wind patterns could change, resulting in climatic changes throughout the world.

SOLUTIONS

He discovery of the ozone depletion problem came as a great surprise. Now, action must be taken to ensure that the ozone layer is not destroyed. Because CFCs are so widespread and used in such a great variety of products, limiting their use is hard. Also, since many products already contain components that use CFCs, it would be difficult if not impossible to eliminate those CFCs already in existence.

The CFC problem may be hard to solve because there are already great quantities of CFCs in the environment. CFCs would remain in the stratosphere for another 100 years even if none were ever produced again.

Despite the difficulties, international action has been taken to limit CFCs. In the Montreal Protocol, 30 nations worldwide agreed to reduce usage of CFCs and encouraged other countries to do so as well. However, many environmentalists felt the treaty did "too little, too late", as theNatural Resources Defence Council put it. The treaty asked for CFC makers to only eliminate half of their CFC production, making some people feel it was inadequate.

By the year 2000, the US and twelve nations in Europe have agreed to ban all use and production of CFCs. This will be highly significant, because these countries produce three quarters of the CFCs in the world. Many other countries have signed treaties and written laws restricting the use of CFCs. Companies are finding substitutes for CFCs, and people in general are becoming more aware of the dangers of ozone depletion.

CHEMICAL PROCESSES

About 90 per cent of the Earth's protective ozone layer resides in the stratosphere between 15km and 50km altitude. Molecular oxygen is broken down in the stratosphere by solar radiation to yield atomic oxygen, which then combines with molecular oxygen to produce ozone. Ozone is destroyed naturally through a series of catalytic cycles involving oxygen, nitrogen, hydrogen and to a lesser extent chlorine and bromine species.

The abundance of stratospheric ozone is therefore chemically controlled by the stratospheric abundances of compounds containing hydrogen, nitrogen, chlorine and bromine. Increases in the abundances of methane and nitrous oxide (sources of hydrogen and nitrogen oxides respectively) thus affect the abundance and distribution of stratospheric ozone.

Stratospheric ozone is also affected by the abundance of carbon dioxide (CO_2), because the rates of the chemical reactions that control the abundance of ozone are temperature-dependent, and the abundance of CO_2 plays a key role in determining the temperature structure of the stratosphere.

During the 1970s and early 1980s, theoretical model calculations focused on predicting the response of stratospheric ozone to changes in chlorine, assuming that the atmospheric abundances of other trace gases remained constant.

However, since the early 1980s, with advances in understanding of trace gas trends and model formulation, model calculations have been used to predict the response of stratospheric ozone to simultaneous increases in chlorine (from chlorofluorocarbons, hydrochlorofluorocarbons, carbon tetrachloride, methylchlo-roform and methyl chloride) and bromine (from halons and methyl bromide), as well as methane, nitrous oxide and carbon dioxide.

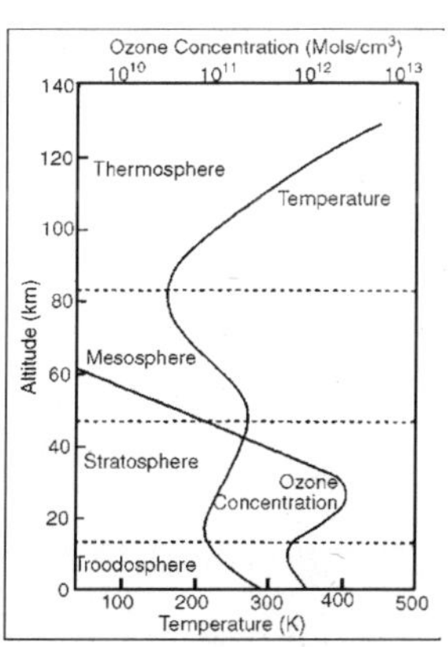

Fig. 1.2: Temperature Profile and Vertical Distribution of Ozone in the Atmosphere

RESPONSES

Policy Responses

Summarizes the main international policy responses in the 1970s and 1980s to the ozone layer issue. Established by UNEP in the 1970s, the Coordinating Committee on the ozone Layer (CCOL) has played a major role in establishing scientific research priorities, and in describing the implications of the latest scientific findings in terms understandable to policy-makers. In 1980, for example, the CCOL reported that continued releases of CFCs would eventually deplete the ozone layer, and that this could have serious impacts on the health of people and the biosphere. Subsequently, the CCOL synthesized the knowledge base to provide the scientific basis for the Montreal Protocol. The policy implications of the scientific findings of the last decade have been profound, leading, for example, to a significant revision of the control measures of the Montreal Protocol.

The most recent scientific findings also have substantial policy implications and could lead to further significant revisions in the control measures of the Montreal Protocol at the Fourth meeting of the Contracting Parties.

Simulation Models in Exploring Policy

Nowadays, simulation models are used more and more to explore policy options for resolving the ozone depletion issue. A particular example, performed, was a sensitivity study of atmospheric chlorine-loading scenarios for a range of global emissions of compounds currently regulated under the Montreal Protocol, and for possible new HCFC substitutes. The results are summarized in Figure below, which shows that:

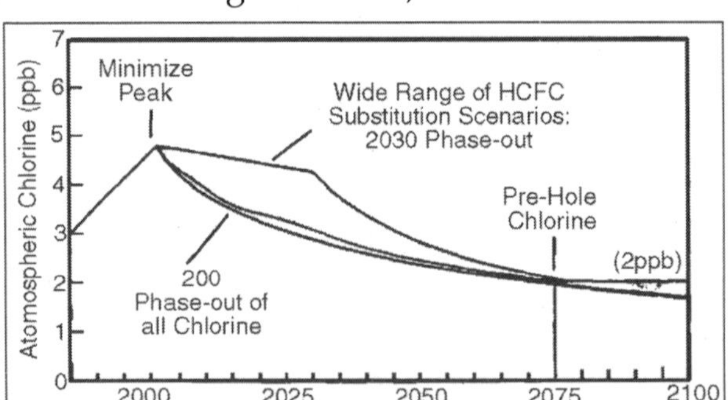

Fig. Impact of Chloroflurocarbon (CFC) and Hydrochlorofluorocarbon (HVFC) Substitution Policies on Atmospheric Abundance of Chlorine (Cl).

- Peak chlorine loading is controlled by the date of the global phase-out (assumed to be the year 2000) for long-lived CFCs. and by emission rates between 1990 and 2000;

- Even with a complete phase-out of all chlorine emissions in 2000, the atmospheric level of chlorine will not decrease to 2 ppbv (the level of chlorine prior to the formation of the Antarctic ozone hole) until about 2075;
- A wide range of HCFC substitution policies can be implemented without delaying recovery of the Antarctic ozone hole, assuming that there is a phase-out of HCFC substitutes before the middle of the next century (the required phase-out date for the HCFCs is dependent upon their lifetime); and
- The additional atmospheric loading of chlorine due to HCFC substitution depends upon the HCFC emission rates and the lifetimes of the HCFCs, but is typically limited to less than 1 ppbv even with significant initial substitution rates and subsequent growth in these emission rates.
- The study concluded that in order to minimize future ozone layer depletion and to facilitate the most rapid elimination of the Antarctic ozone hole (*i.e.* by about 2075):
- Production and emission of long-lived halocarbons (CFCs and CCl4) should be phased out as soon as possible, with the greatest possible global compliance:
- Halocarbon substitutes should be phased out sometime in the middle of the next century:
- Atmospheric concentrations of bromine should be stabilized at current levels or below.
- Peak chlorine loadings should be reduced, most practicably by significant cutbacks in the emissions of abundant, short-lived halocarbons, particularly methyl chloroform.

CONCLUDING REMARKS

Depletion of the stratospheric ozone layer by halogenated chemicals is a global problem. While the issue was primarily a matter of scientific curiosity during the 1970s and early 1980s, it has now become an urgent policy question for governments in both developed and developing countries. The weight of scientific evidence strongly indicates that chlorinated and brominated chemicals (largely man-made) are primarily responsible for the substantial decreases in stratospheric ozone (greater than 50 per cent in the total column) that occur over Antarctica every Southern Hemisphere springtime.

In addition, satellite data supported by ground-based observations have demonstrated that ozone has been decreasing since the late 1960s at middle and high latitudes in both the northern and southern hemispheres, and that this decrease cannot be explained by known natural processes. Indeed, the conclusion of the 1991 International Scientific Assessment was that 'the weight of evidence suggests that the observed middle- and high-latitude ozone losses are largely due to chlorine and bromine'. The current and historic use of CFCs and other such chemicals in developed countries is the primary cause of the problem. However, adequate protection of the ozone layer will require a full partnership between the developed countries, whose industries and consumers have caused the problem, and the developing countries, whose people wish to use these chemicals in refrigeration and other ways.

As in the case of atmospheric concentrations of lead, which are now declining, even in Greenland, as a result of tighter emission controls, the stratosphere would gradually recover if emissions of CFCs and similar chemicals were to cease.

Thus, the most urgent task is to strengthen the Montreal Protocol and to develop an international method of monitoring national Programmemes to control CFCs and related gases. However scientific research and monitoring should continue into the next century to provide a basis for better policies.

GREENHOUSE EFFECT

The greenhouse effect is a naturally occurring process that aids in heating the Earth's surface and atmosphere. It results from the fact that certain atmospheric gases, such as carbon dioxide, *water vapour*, and methane, are able to change the energy balance of the planet by absorbing longwave radiation emitted from the Earth's surface. Without the greenhouse effect life on this planet would probably not exist as the average temperature of the Earth would be a chilly -18° Celsius, rather than the present 15° Celsius.

As energy from the Sun passes through the atmosphere a number of things take place. A portion of the energy (26% globally) is reflected or scattered back to space by clouds and other atmospheric particles. About 19% of the energy available is absorbed by clouds, gases (like ozone), and particles in the atmosphere. Of the remaining 55% of the solar energy passing through the Earth's atmosphere, 4% is reflected from the surface back to space. On average, about 51% of the Sun's radiation reaches the surface. This energy is then used in a number of processes, including the heating of the ground surface; the melting of ice and snow and the evaporation of water; and plant photosynthesis.

The heating of the ground by sunlight causes the Earth's surface to become a radiator of energy in the longwave band (sometimes called infrared radiation).

Absorption of longwave radiation by the atmosphere causes additional heat energy to be added to the Earth's atmospheric system. The now warmer atmospheric greenhouse gas molecules begin radiating longwave energy in all directions. Over 90% of this emission of longwave energy is directed back to the Earth's surface where it once again is absorbed by the surface. The heating of the ground by the longwave radiation causes the ground surface to once again radiate, repeating the cycle described above, again and again, until no more longwave is available for absorption.

The amount of heat energy added to the atmosphere by the greenhouse effect is controlled by the concentration of greenhouse gases in the Earth's atmosphere. All of the major greenhouse gases have increased in concentration since the beginning of the Industrial Revolution. As a result of these higher concentrations, scientists predict that the greenhouse effect will be enhanced and the Earth's climate will become warmer.

Predicting the amount of warming is accomplished by computer modeling. Computer models suggest that a doubling of the concentration of the main greenhouse gas, carbon dioxide, may raise the average global temperature between 1 and 3° Celsius. However, the numeric equations of computer models do not accurately simulate the effects of a number of possible negative feedbacks. For example, many of the models cannot properly simulate the negative effects that increased cloud cover would have on the radiation balance of a warmer Earth.

Increasing the Earth's temperature would cause the oceans to evaporate greater amounts of water, causing the atmosphere to become cloudier. These extra clouds would then reflect a greater proportion of the Sun's energy back to space reducing the amount of solar radiation absorbed by the atmosphere and the Earth's surface. With less solar energy being absorbed at the surface, the effects of an enhanced greenhouse effect may be counteracted. Artificially created chlorofluorocarbons are the strongest greenhouse gas per molecule. However, low concentrations in the atmosphere reduce their overall importance in the enhancement of the greenhouse effect. Current measurements in the atmosphere indicate that the concentration of these chemicals may soon begin declining because of reduced emissions.

Reports of the development of ozone holes over the North and South Poles and a general decline in global stratospheric ozone levels over the last two decades has caused many nations to cutback on their production and use of these chemicals. In 1987, the signing of the Montreal Protocol agreement by forty-six nations established an immediate timetable for the global reduction of chlorofluorocarbons production and use.

Since 1750, methane concentrations in the atmosphere have increased by more than 150%. The primary sources for the additional methane added to the atmosphere (in order of importance) are rice cultivation, domestic grazing animals, termites, landfills, coal mining, and oil and gas extraction. Anaerobic conditions associated with rice paddy flooding results in the formation of methane gas. However, an accurate estimate of how much methane is being produced from rice paddies has been difficult to obtain. More than 60% of all rice paddies are found in India and China where scientific data concerning emission rates are unavailable.

Nevertheless, scientists believe that the contribution of rice paddies is large because this form of crop production has more than doubled since 1950. Grazing animals release methane to the environment as a result of herbaceous digestion. Some researchers believe the addition of methane from this source has more than quadrupled over the last century.

Termites also release methane through similar processes. Land-use change in the tropics, due to deforestation, ranching, and farming, may be causing termite numbers to expand. If this assumption is correct, the contribution from these insects may be important. Methane is also released from landfills, coal mines, and gas and oil drilling. Landfills produce methane as organic wastes decompose over time. Coal, oil, and natural gas deposits release methane to the atmosphere when these deposits are excavated or drilled.

The average concentration of nitrous oxide in the atmosphere is now increasing at a rate of 0.2 to 0.3% per year. Sources for this increase include land-use conversion; fossil fuel combustion; biomass burning; and soil fertilization. Most of the nitrous oxide added to the atmosphere each year comes from deforestation and the conversion of forest, savanna and grassland ecosystems into agricultural fields and rangeland.

Both of these processes reduce the amount of nitrogen stored in living vegetation and soil through the decomposition of organic matter. Nitrous oxide is also released into the atmosphere when fossil fuels and biomass are burned. However, the combined contribution of these sources to the increase of this gas in the atmosphere is thought to be minor.

The use of nitrate and ammonium fertilizers to enhance plant growth is another source of nitrous oxide. Accurate measurements of how much nitrous oxide is being released from fertilization have been difficult to obtain. Estimates suggest that the contribution from this source may represent from 50% to 0.2% of nitrous oxide added to the atmosphere annually. Ozone's role in the enhancement of the greenhouse effect has been difficult to determine scientifically.

Accurate measurements of past long-term (more than 25 years in the past) levels of this gas in the atmosphere are currently unavailable. Concentrations of ozone gas are found in two different regions of the Earth's atmosphere. The majority of the ozone (about 97%) found in the atmosphere is localized in the stratosphere at an altitude of 15 to 55 Kilometres above the Earth's surface.

In recent years, the concentration of the stratospheric ozone has been decreasing because of the buildup of chlorofluorocarbons in the atmosphere. Since the late 1970s, scientists have discovered that total column ozone amounts over Antarctica in the springtime have decreased by as much as 70%. Satellite measurements have indicated that the zone from 65° North to 65° South latitude has had a 3% decrease in stratospheric ozone since 1978. Ozone is also highly concentrated at the Earth's surface.

Most of this ozone is created as an artificial by product of photochemical smog. In summary, the greenhouse effect causes the atmosphere to trap more Effect of Environment 263 heat energy at the Earth's surface and within the atmosphere by absorbing and re-emitting longwave energy. Of the longwave energy emitted back to space, 90% is intercepted and absorbed by greenhouse gases.

In the last few centuries, the activities of humans have directly or indirectly caused the concentration of the major greenhouse gases to increase. Scientists predict that this increase mayenhance the greenhouse effect making the planet warmer. Some experts estimate that the Earth's average global temperature has already increased by 0.3 to 0.6° Celsius, since the beginning of this century, because of this enhancement. Predictions of future climates indicate that by the middle of the next century the Earth's global temperature may be 1 to 3° Celsius higher than today.

BASIC MECHANISM

The Earth receives energy from the Sun mostly in the form of visible light. The bulk of this energy is not absorbed by the atmosphere since the atmosphere is transparent to visible light. 50% of the sun's energy reaches the Earth which is absorbed by the surface as heat.

Because of its temperature, the Earth's surface radiates energy in infrared range.The Greenhouse gases are not transparent to infrared radiation so they absorb infrared radiation. Infrared radiation is absorbed from all directions and is passed as heat to all gases in the atmosphere.

The atmosphere also radiates in the infrared range (because of its temperature, in the same way the Earth's surface does) and does so in all directions. The surface and lower atmosphere are warmed because of the greenhouse gases and makes our life on earth possible.

DETAILED EXPLANATION

The Earth receives energy from the Sun in the form of radiation. Most of the energy is in visible wavelengths and in infrared wavelengths that are near the visible range (often called "near infrared"). The Earth reflects about 30% of the incoming solar radiation. The remaining 70% is absorbed, warming the land, atmosphere and ocean.

For the Earth's temperature to be in steady state so that the Earth does not rapidly heat or cool, this absorbed solar radiation must be very closely balanced by energy radiated back to space in the infrared wavelengths. Since the intensity of infrared radiation increases with increasing temperature, one can think of the Earth's temperature as being determined by the infrared flux needed to balance the absorbed solar flux.

The visible solar radiation mostly heats the surface, not the atmosphere, whereas most of the infrared radiation escaping to space is emitted from the upper atmosphere, not the surface. The infrared photons emitted by the surface are mostly absorbed in the atmosphere by greenhouse gases and clouds and do not escape directly to space.

The reason this warms the surface is most easily understood by starting with a simplified model of a purely radiative greenhouse effect that ignores energy transfer in the atmosphere by convection (sensible heat transport, Sensible heat flux) and by the evaporation and condensation of water vapour (latent heat transport, Latent heat flux). In this purely radiative case, one can think of the atmosphere as emitting infrared radiation both upwards and downwards. The upward infrared flux emitted by the surface must balance not only the absorbed solar flux but also this downward infrared flux emitted by the atmosphere. The surface temperature will rise until it generates thermal radiation equivalent to the sum of the incoming solar and infrared radiation.

A more realistic picture taking into account the convective and latent heat fluxes is somewhat more complex. But the following simple model captures the essence. The starting point is to note that the opacity of the atmosphere to infrared radiation determines the height in the atmosphere from which most of the photons are emitted into space. If the atmosphere is more opaque, the typical photon escaping to space will be emitted from higher in the atmosphere, because one then has to go to higher altitudes to see out to space in the infrared. Since the emission of infrared radiation is a function of temperature, it is the temperature of the atmosphere at this emission level that is effectively determined by the requirement that the emitted flux balance the absorbed solar flux.

But the temperature of the atmosphere generally decreases with height above the surface, at a rate of roughly 6.5 °C per Kilometre on average, until one reaches the stratosphere 10 15 km above the surface.

(Most infrared photons escaping to space are emitted by the troposphere, the region bounded by the surface and the stratosphere, so we can ignore the stratosphere in this simple picture.) A very simple model, but one that proves to be remarkably useful, involves the assumption that this temperature profile is simply fixed, by the non-radiative energy fluxes.

Given the temperature at the emission level of the infrared flux escaping to space, one then computes the surface temperature by increasing temperature at the rate of 6.5 °C per Kilometre, the environmental lapse rate, until one reaches the surface.

The more opaque the atmosphere, and the higher the emission level of the escaping infrared radiation, the warmer the surface, since one then needs to follow this lapse rate over a larger distance in the vertical. While less intuitive than the purely radiative greenhouse effect, this less familiar radiative-convective picture is the starting point for most discussions of the greenhouse effect in the climate modeling literature.

LA NINO

The term El Niño (Spanish for "the little boy" or "the Christ Child") was originally used by fishermen along the coasts of Ecuador and Peru to refer to above-normal sea-surface temperatures that typically appear around Christmas time in the eastern Pacific Ocean and last for several months. But El Niño's effects are not limited to Peru and Ecuador.

WORKING OF EL NIÑO

In normal years, winds tend to blow from east to west across the waters of the tropical Pacific. The easterly winds (remember that wind directions refer to the direction FROM which the wind comes: an easterly trade wind comes from the EAST and blows WEST) push the surface waters westward across the ocean. In turn, this causes deeper, colder waters to rise to the surface. This "upwelling" of deep ocean waters brings with it nutrients that otherwise would remain near the bottom.

During El Nino years, the winds weaken, reducing or even choking off the upwelling of deep water. The consequent warming of the ocean surface further weakens the winds and strengthens El Niño.

As the Pacific continues to heat up, the warmer waters shift eastward, and so do the clouds and thunderstorms that produce heavy rainfall along the equator. This results in changes in jet streams (the winds high aloft), which lead to dry conditions in Indonesia and Australia, and floods in Peru and Ecuador. El Niño events occur on average every 3 to 5 years, though there can be periods of up to a decade without an El Niño.

EL NIÑO'S EFFECTS

The 1982-83 El Niño was unusually strong. In Ecuador and northern Peru, up to 100 inches of rain fell during a six-month period, transforming the coastal desert into grassland dotted with lakes.

Abnormal wind patterns also caused the monsoon rains to fall over the central Pacific instead of on the western shores, which led to droughts and 266 Encyclopaedia of Environmental Science disastrous forest fires in Indonesia and Australia. Overall, the loss to the global economy as a result of the El Niño amounted to more than $8 billion.

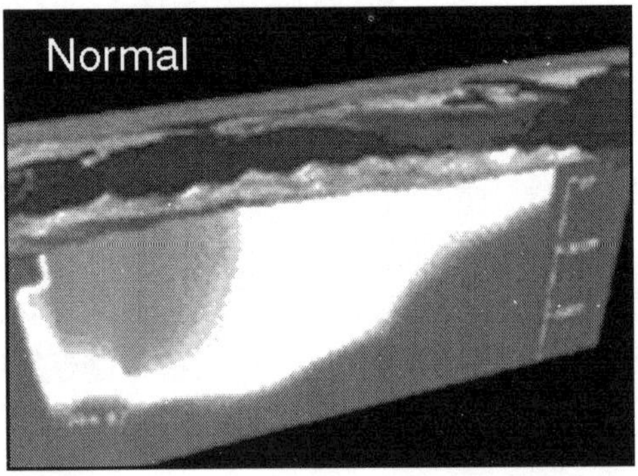

Likewise, the winter of 1997-1998 was marked by a record-breaking El Niño event. The result was unusual weather in parts of the world, including the U.S. Severe weather events included flooding in the southeastern United States, major storms in the Northeast, and flooding in California.

❒

2

Environmental Microbiology

DEFINITION

The study of how microorganisms affect the earth and its atmosphere is called environmental microbiology or microbial ecology. Environmental microbiology is the study of microbes in the environment and their interactions with each other. Microbes are the tiniest creatures on Earth, yet despite their small size, they have a have a huge impact on us and on our environment.

Most types of microbes remain unknown. It is estimated that we know fewer than 1% of the microbial species on Earth. Yet microbes surround us everywhere—air, water, soil. An average gram of soil contains one billion (1,000,000,000) microbes representing probably several thousand species. Environmental microbiology is the study of the relationships that exist between microorganisms and the environment.

Environmental microbiology is the relationship of microorganisms with themselves and with their surroundings. Microorganisms have special impact on the whole biosphere. They are the backbone of ecosystems of the zones where light cannot approach. In such zones, chemosynthetic bacteria are present which provide energy and carbon to the other organisms there. Some microbes are decomposers which have ability to recycle the nutrients. So, microbes have a special role in biogeochemical cycles. Microbes, especially bacteria, are of great importance in the sense that their symbiotic relationship (either positive or negative) have special effects on the ecosystem.

Bacteria and their enzymes, along with some fungi and critical nutrient additives are cost effective agents for in-situ remediation of hazardous wastes and subsurface pollution in soils, sediments and wastewaters. The ability of each bacteria strain to degrade toxic waste depends on the nature of each contaminant. Since most sites are typically comprised of multiple pollutant types, the most effective approach to bioremediation is to use a mixture of bacterial species/strains, each specific to the degradation of one or more types of contaminants. It is critical to monitor the composition of the indigenous and added bacterial consortium in order to evaluate the activity level of the bacteria, and to permit modifications of the nutrients and other conditions for optimizing the bioremediation process.

OVERVIEW

The pervasive and manifold positive and negative impacts of microbial activities on human and environmental health, and human activities, require that efforts to improve the human condition include efforts to modulate microbial activities. Attainment of this goal will be greatly facilitated by acquisition of a fundamental understanding of microbial functioning in their natural settings, and of parameters that regulate such activities. Our current fragmentary knowledge of microbial processes is based largely on information obtained from individual strains of microbes studied in the Labouratory as pure, exponentially-growing cultures in nutrient-rich, homogeneous aqueous solutions.

Sewage treatment is the process that removes the majority of the contaminants from waste-water or sewage and produces both a liquid effluent suitable for disposal to the natural environment and a sludge. To be effective, sewage must be conveyed to a treatment plant by appropriate pipes and infrastructure and the process itself must be subject to regulation and controls. Sewage is the liquid waste from toilets, baths, showers, kitchens, etc. that is disposed of via sewers. In many areas sewage also includes some liquid waste from industry and commerce. In the UK, the waste from toilets is termed foul waste, the waste from items such as basins, baths, kitchens is termed sullage water, and the industrial and commercial waste is termed trade waste.

The division of household water drains into Greywater and Black water is becoming more common in the United States, with greywater being permitted to be used for watering plants or recycled for flushing toilets. Much sewage also includes some surface water from roofs or hard-standing areas. Municipal wastewater therefore includes residential, commercial, and industrial liquid waste discharges, and may include stormwater run-off.

Sewerage systems that transport liquid waste discharges and stormwater together to a common treatment facility are called combined sewer systems. The construction of combined sewers is a less common practice in the U.S. and Canada than in the past and is no longer accepted within Building Regulations in the UK. Instead, liquid waste and stormwater are collected and conveyed in separate sewer systems, referred to as sanitary sewers and storm sewers in the U.S. and as foul sewers and surface water sewers in the UK. Overflows from foul sewers designed to relieve pressure from heavy rainfall are termed storm sewers or combined sewer overflows.

As rainfall runs over the surface of roofs and the ground, it may pick up various contaminants including soil particles (sediment), metals, organic compounds, animal waste, and oil and grease. Some jurisdictions, such as certain communities located in southern California, require stormwater to receive some level of treatment before being discharged to the environment. Examples of treatment processes used for stormwater include sedimentation basins, wetlands, and vortex separators (to remove coarse solids).

"Conventional wastewater treatment" is the process used in modern wastewater treatment plants that removes the majority of the contaminants from wastewater. This produces a liquid effluent suitable for disposal to the natural environment and also produces a sludge.

The treatment process typically involves the following three stages: Primary treatment Primary treatment is to reduce oils, grease, fats, sand, grit, and coarse (settleable) solids.

Grit removal This stage typically includes a grit channel where the velocity of the incoming wastewater is carefully controlled to allow grit and stones to settle but still maintain all organic material within the flow. Grit and stones need to be removed early in the process to avoid damage to pumps and equipment in the remaining treatment stages.

Screening or maceration The grit free liquid is then passed through fixed or rotating screens to remove larger material such as rags. Screenings are collected and may be returned to the sludge treatment plant or may be disposed of off site by landfilling or incineration. Maceration, in which solids are cut into small particles through the use of rotating knife edges mounted on a revolving cylinder, is used in plants that are able to process this particulate waste. Macerators are, however, more expensive to maintain and are less reliable than physical screens.

SEDIMENTATION

Primary sedimentation tank at a rural treatment plantIn almost all plants there is a sedimentation stage where the sewage is allowed to stand in large tanks so that faecal solids can settle and floating material such as grease and plastics can rise to the surface and be skimmed off. The main purpose of the primary stage is to produce a generally homogeneous liquid capable of being treated biologically together with a sludge that can be separately treated or processed. Primary settlement tanks are usually equipped with mechanically driven scrapers that continually drive the collected sludge towards a hopper in the base of the tank from where it can be pumped to further sludge treatment stages.

Secondary treatment Secondary treatment is designed to substantially degrade the biological content of the sewage. The majority of municipal and industrial plants treat the settled sewage liquor using aerobic biological processes. For this to be effective, the biota require both oxygen and a substrate on which to live. There are number of ways in which this is done. In all these methods, the bacteria and protozoa consume biodegradable soluble organic contaminants (*e.g.* sugars, fats, organic short-chain carbon molecules, etc.) and bind much of the less soluble fractions into floc particles.

Roughing Filters Roughing filters are intended to treat particularly strong or variable organic loads. They are typically tall, columnar filters filled with open synthetic filter media to which sewage is applied at a relatively high rate. The design of the filters allows high hydraulic loading and a high flow- through of air. The resultant liquor is usually within the normal range for conventional treatment processes.

Activated sludge Activated sludge plants use a variety of mechanisms and processes to use dissolved oxygen to generate a biological floc that substantially removes organic material. It also traps particulate material and can, under ideal conditions, convert ammonia to nitrite or nitrate.

FILTER BEDS

Trickling filter bed using plastic mediaIn older plants and plants receiving more variable loads, trickling filter beds are used where the settled sewage liquor is spread onto the surface of a deep bed made up of coke (carbonised coal), rocks or specially fabricated plastic media with high surface areas. The liquor is distributed through perforated rotating arms radiating from a central pivot.

The distributed liquor trickles through this bed and is collected in drains at the base. These drains also provide a source of air which percolates up through the bed, keeping it aerobic. Biological film comprising of bacteria, protozoa and fungi forms on all the available surfaces and this provides the required biological treatment capability to effect the reduction in organic content. Rotating Plates and Spirals In some smaller plants slowly revolving plates or spirals are used which are partially submerged in the liquor. A biotic floc is created which provides the required substrate.

Secondary sedimentation The final step in the secondary treatment stage is to settle out the biological floc or filter material and produce an effluent with very low levels of organic material and suspended matter.

Secondary Sedimentation tank at a rural treatment plantTertiary treatment Tertiary treatment provides a final stage to raise the effluent quality to the standard required before it is discharged to the receiving environment (sea, river, lake, ground, etc.) More than one tertiary treatment process may be used at any treament plant. If disinfection is practiced, it is always the final process.

Effluent polishing Sand filtration Sand filtration removes much of the residual suspended matter. Lagooning Lagoons provides settlement and further biological improvement through storage in large man-made ponds or lagoons.

Constructed wetlands These include engineered reed beds and a range of similar methodologies, all of which provide a high degree of aerobic biological improvement and can often be used instead of secondary treatment for small communities.

Nutrient removal Wastewater may also contain high levels of nutrients (nitrogen and phosphorus) that in certain forms may be toxic to fish and invertebrates at very low concentrations(*e.g.* ammonia) or that can create nuisance conditions in the receiving environment (*e.g.* weed or algal growth). Although the growth of weeds and algae may seem to be primarily an aesthetic issue, algae can produce toxins, and in dying their decay and consumption by bacteria in the environment can result in the depletion of oxygen in the water and the possible consequential suffocation of fish. Where receiving rivers discharge to lakes or shallow seas, the added nutrients can cause severe and sometimes irreversible eutrophication with the loss of many sensitive clean water species. The removal of nitrogen and/or phosphorus from wastewater can be achieved either biologically or by chemical precipitation treatment processes.

Nitrogen removal Biological treatment of nitrogen generally involves creating conditions within the treatment process for bacteria to convert the ammonia to nitrate, and then allowing other bacteria to reducing the nitrate to nitrogen gas, which is released to the atmosphere. Sand filters, lagooning and the use of reed beds can all be used to reduce nitrogen. Sometimes the conversion of toxic ammonia to nitrate alone is referred to as tertiary treatment.

Phosphorous removal The biological treatment of wastewater to remove phosphorus also involves the design and creation of specific environmental conditions within a treatment plant to enable specific bacteria to bio-accumulate large quantities of phosphorus. When the bacteria containing the phosphorus are removed, the resulting bacterial biosolids often have a high fertilizer value. Phosphorus can also be removed by chemical precipitation using (commonly) salts of iron (*i.e.* ferric chloride) or aluminum (*i.e.* alum). The resulting chemical sludge, however, is difficult to dispose of, and the use of chemicals in the treatment process is expensive and makes operation difficult and often messy.

Disinfection Disinfecting substantially reduces the numbers of living organisms in the water to be discharged. The effectiveness of disinfection depends on the quality of the water being treated, the type of disinfection being used, the application rate, the contact time and environmental variables. Turbid water will be treated less successfully since solid matter can shield organisms, especially from Ultraviolet light or if contact times are low. Generally, short contact times, low doses and high flows all mitigate against effective disinfection. Common methods of disinfection include ozone, chlorine, or UV light. Chloramine, which is used for drinking water, is not used in waste water treatment because of its persistence.

A highly efficient dye-deColourizing bacterium, strain S12(T), was isolated from activated sludge of a textile-printing waste-water treatment plant in Guangzhou, China. The cells were Gram-negative and motile by means of a single polar flagellum. The strain was capable of anaerobic growth either by fermentation of glucose or by anaerobic respiration and utilized a variety of electron acceptors, including nitrate, iron oxide and thiosulfate. The physiological properties, tested by using the Biolog GN2 system, were similar to those of the genus of Shewanella. Analysis of the nearly complete 16S rRNA gene sequence of strain S12(T) showed the highest similarity (98 and 97%, respectively) to Shewanella baltica and Shewanella putrefaciens. However, the level of gyrB similarity between strain S12(T) and S. putrefaciens was 87%. DNA from strain S12(T) showed 41.8 and 41.9% DNA relatedness, respectively, to the DNA of S. baltica DSM 9439(T) and S. putrefaciens DSM 6067(T). The DNA G+C content of strain S12(T) was 49.3 mol%. The predominant menaquinone was MK-7 and the predominant ubiquinones were Q-7 and Q-The dominant fatty acids were 15: 0, 16: 0, iso-15: 0 and 16: 1omega7c, similar to the profiles of other Shewanella species. On the basis of its physiological and molecular properties, strain S12(T) appears to represent a novel species of the genus Shewanella, for which the name Shewanella deColourationis sp. nov. is proposed. The type strain is S12(T) (=CCTCC M 203093(T)=IAM 15094(T)).

Hydrogenophaga palleronii (98.5%) and Hydrogen-ophaga taeniospiralis (98.0%). The fatty acid patterns and substrate-utilization profiles displayed similarity to the those of the five Hydrogenophaga species with validly published names, although clear differentiating characteristics were also observed. The two strains showed DNA-DNA hybridization values of 51% with respect to each other. No close similarities to any other Hydrogenophaga species were detected in hybridization experiments with the genomic DNAs. On the basis of these results, two novel Hydrogenophaga species, Hydrogenophaga defluvii sp. nov. and Hydrogenophaga atypica sp. nov. are proposed, with BSB 9.5(T) (=DSM 15341(T)=CIP 108119(T)) and BSB 41.8(T) (=DSM 15342(T)=CIP 108118(T)) as the respective type strains.

Effect of ethoxylate number and alkyl chain length on the pathway and kinetics of linear alcohol ethoxylate biodegradation in activated sludge; Itrich NR et al.; Batch activated-sludge die-away studies were conducted with various pure homologs to determine the effect of ethoxylate number and alkyl chain length on the kinetics of primary and ultimate biodegradation of linear alcohol ethoxylates. The 14C-(ethoxylate) homologs $C_{14}E_1$, $C_{14}E_3$, $C_{14}E_6$, and $C_{14}E_9$ were used to investigate the effect of ethoxylate number, and 14C-(ethoxylate) homologs $C_{12}E_6$, $C_{14}E_6$, and $C_{16}E_6$ were used to examine the effect of chain length. Activated sludge was dosed with a trace concentration (0.2 microM) of each homolog, and the disappearance of parent, formation of metabolites, production of $14CO_2$, and uptake into solids were monitored with time. Ethoxylate number had little effect on the first-order decay rates for primary biodegradation, which ranged from 61 to 78 h(−1).

However, alkyl chain length had a larger effect, with the C16 chain-length homolog exhibiting a slower rate of parent decay (18 h(−1)) compared to its corresponding C12 and C14 homologs (61-69 h(1)). Ethoxylate number affected the mechanism of biodegradation, with fission of the central ether bond to yield the corresponding fatty alcohol and (poly) ethylene glycol group increasing in dominance with increasing ethoxylate number. Based upon the measured rates of primary biodegradation, removal of parent during activated- sludge treatment was predicted to range between 99.7 and 99.8% for all homologs except C16E6, which had a predicted removal of 98.9%.

Based upon the measured rates of ultimate bio-degradation, removal of ethoxylate-containing metabolites was predicted to exceed 83% for all homologs. These predictions corresponded closely with previously published removal measurements in Labouratory continuous activated-sludge systems and actual treatment plants.

It was found that the estimated theoretical maximum growth yield of aerobic granules was as low as 0.2 g biomass g(-1) chemical oxygen demand (COD) and 10-40% of input substrate-COD was consumed through the maintenance metabolism, while experimental results further showed that the unit oxygen uptake by aerobic granules was 0.68 g oxygen g(–1) COD, which was much higher than that reported in activated sludge processes. Based on the growth yield and unit oxygen uptake determined, an oxidative assimilation equation of acetate-fed aerobic granules was derived; and this was confirmed by respirometric tests. In aerobic granular culture, about 74% of the input substrate-carbon was converted to carbon dioxide. The growth yield of aerobic granules was three times lower than that of activated sludge. It is likely that high carbon dioxide production is the main cause of the low growth yield of aerobic granules, indicating a possible energy uncoupling in aerobic

GRANULAR CULTURE

Comparing activated sludge and aerobic granules as microbial inocula for phenol biodegradation; Tay ST et al.; Activated sludge and acetate-fed granules were used as microbial inocula to start up two sequencing batch reactors (R_1, R_2) for phenol biodegradation. The reactors were operated in 4-h cycles at a phenol loading of 1.8 kg m(–3) day(–1). The biomass in R1 failed to remove phenol and completely washed out after 4 days. R_2 experienced initial difficulty in removing phenol, but the biomass acclimated quickly and effluent phenol concentrations declined to 0.3 mg 1(–1) from day 3.

The acetate-fed granules were covered with bacterial rods, but filamentous bacteria with sheaths, presumably to shield against toxicity, quickly emerged as the dominant morphotype upon phenol exposure. Bacterial adaptation to phenol also took the form of modifications in enzyme activity and increased production of extracellular polymers. 16S rRNA gene fingerprints revealed a slight decrease in bacterial diversity from day 0 to day 3 in R1, prior to process failure.In R_2, a clear shift in community structure was observed as the seed evolved into phenol-degrading granules without losing species-richness. The results highlight the effectiveness of granules over activated sludge as seed for reactors treating toxic wastewaters.

Processes for managing pathogens; Godfree A et al.; Wastewater contains human, animal, and plant pathogens capable of causing viral, bacterial, or parasitic infections. There are several routes whereby sewage pathogens may affect human health, including direct contact, contamination of food crops, zoonoses, and vectors. The range and numbers of pathogens in municipal wastewater vary with the level of endemic disease in the community, discharges from commercial activities, and seasonal factors.

Regulations to control pathogen risk in the United States and Europe arising from land application of biosolids are based on the concept of multiple barriers to the prevention of transmission.

The barriers are:

- Treatment to reduce pathogen content and vector attraction.
- Restrictions on crops grown on land to which biosolids have been applied.
- Minimum intervals following application and grazing or harvesting.

Wastewater treatment reduces number of pathogens in the wastewater by concentrating them with the solids in the sludge. Although some treatment processes are designed specifically to inactivate pathogens, many are not, and the actual mechanisms of microbial inactivation are not fully understood for all processes. Vector attraction is reduced by stabilization (reduction of readily biodegradable material) and/or incorporation immediately following application.

Concerns about health risks have renewed interest in the effects of treatment (on pathogens) and advanced treatment methods, and work performed in the United States suggests that Class A pathogen reduction can be achieved less expensively than previously thought. Effective pathogen risk management requires control to the complete chain of sludge treatment, biosolids handling and application, and post-application activities. This may be achieved by adherence to quality management systems based on hazard analysis critical control point (HACCP) principles.

Evaluation of microscopic techniques (epifluorescence microscopy, CLSM, TPE-LSM) as a basis for the quantitative image analysis of activated sludge; Lopez C et al.; Microscopic techniques ranging from epifluorescence microscopy to confocal laser scanning microscopy (CLSM) and two photon excitation laser scanning microscopy (TPE-LSM) combined with fluorescent stains can help to evaluate complex microbial aggregates such as activated sludge flocs. To determine the application limits of these microscopic techniques, activated sludge samples from three different sources were evaluated after staining with a fluorescent viability indicator (Baclight Bacterial Viability Kit, Molecular Probes).

Image analysis routines were developed to quantify overall amounts of red and green stained cells, location of stained cells within the flocs, and the spatial organization in clusters and filaments. It was found that the selection of the appropriate microscopic technique depends strongly on the type of microbial aggregates being Analysed.

For flocs with high cell density, the use of TPE-LSM is preferred, since it provides a clearer image of the internal structure of the aggregate.

their natural host cells, and lend further support to suggested implications of environmental chlamydiae for public health.

Fenton peroxidation improves the drying performance of waste activated sludge; Dewil R et al.; Advanced sludge treatment processes (AST) reduce the amount of sludge produced and improve the dewaterability, thus probably also affecting the heat transfer properties and the drying characteristics of the sludge. This chapter studies the influence of the Fenton peroxidation on the thermal conductivity of the sludge.

Results demonstrate that the Fenton's peroxidation positively influences the sludge cake consistency and hence enhances the mechanical dewaterability and the drying characteristics of the dewatered sludge. For the two sludges used in this study, *i.e.* obtained from the wastewater treatment plants (WWTP) of Tienen and Sint-Niklaas - the dry solids content of the mechanically dewatered sludge increased from 22.5% to 40.3% and from 18.7% to 35.2%, respectively. The effective thermal conductivity k(e) of the untreated and the peroxidized sludges is measured and used to determine the heat transfer coefficient h(s).

An average improvement for k(e) of 16.7% (Tienen) and 5.8% (Sint-Niklaas) was observed. Consequently the value of h(s) increased with 15.6% (Tienen) and 5.0% (Sint-Niklaas). This increased heat transfer coefficient in combination with the increased dewaterability has direct implications on the design of sludge dryers. A plate-to-plate calculation of a multiple hearth dryer illustrates that the number of plates required to dry the peroxidized sludge to 90% DS is less than half the number of plates needed to dry untreated sludge. This results in reduced dryer dimensions or a higher capacity for an existing dryer of given dimensions.

Discrepancies in the widely applied GAM42a fluorescence in situ hybridisation probe for Gammaproteobacteria; Siyambalapitiya N et al.; A bacterial culture collection of 104 strains was obtained from an activated sludge wastewater treatment plant to pursue studies into microbial flocculation. Characterisation of the culture collection using a polyphasic approach indicated seven isolates, phylogenetically affiliated with the deep-branching Xanthomonas group of the class Gammaproteobacteria, were unable to hybridise the GAM42a fluorescence in situ hybridisation (FISH) probe for Gammaproteobacteria.

The sequence of the GAM42a probe target region in the 23S rRNA gene of these isolates was determined to have mismatches to GAM42a. Probes perfectly targeting the mismatches (GAM42a_T1038_G1031, and GAM42a_T1038 and GAM42a_A1041_A1040) were synthesised, and used in conjunction with GAM42a in FISH to study the Gammaproteobacteria community structure in one full-scale activated sludge plant. Several bacteria in the activated sludge biomass bound the modified probes demonstrating their presence and the fact that these Gammaproteobacteria have been overlooked in community structure analyses of activated sludge.

Toxicity of metals and organic chemicals evaluated with bioluminescence assays; Ren S et al.; The development of a bioluminescent sensor organism (Shk1) that was created for assessing wastewater toxicity was reported several years ago. In order to establish a test battery to better characterize wastewater toxicity, additional luminescent sensor organisms were later created. The present study focused on one promising candidate (PM6), a Pseudomonas spp. strain, because of its high level of luminescence compared to that of other newly created organisms. Using a batch toxicity testing protocol, the toxicity of 7 metals and 25 organic compounds was evaluated with the PM6 and Shk1 assays. Results indicated that the toxicity data of the PM6 and the Shk1 assays were correlated, and no assay appeared to be particularly more sensitive to a group of toxicants than the other assay. The results of the PM6 and Shk1 assays were further evaluated by comparing with the results of the Vibrio fischeri luminescence inhibition assay and activated sludge inhibition assays. Data suggested that PM6 and Shk1 more closely represented activated sludge organisms than V. fischeri. The suitability of using PM6 and Shk1 for assessing wastewater toxicity on activated sludge, both individually and in a test battery, was discussed.

Aerobic biodegradation of organotin compounds in activated sludge batch reactors; Stasinakis AS et al.; The biodegradation Behaviour of four organotin (OT) compounds, namely tributyltin (TBT), dibutyltin (DBT), monobutyltin (MBT) and triphenyltin (TPhT), was studied in lab-scale activated sludge batch reactors. The activated sludge was spiked with the OT compounds at a level of 100mugl(–1) as Sn.

Determination of the OT compounds by GC-FPD after ethylation in the dissolved and particulate phase revealed that 24h after the start of the experiments, almost the total of OT compounds has been removed from the dissolved phase and is associated with the suspended solids. Calculation of mass balance in batch reactors showed that OT compounds biodegradation was performed via a sequential dealkylation process.

Removals due to biodegradation were differentiated according to the parent compound. In experiments with non-acclimatized biomass, a percentage of 27.1, 8.3, 73.8 and 51.3 was still present as TBT, DBT, MBT and TPhT, respectively, at the end of the experiment (18th day). Half-lives (t(1/2)) of 10.2 and 5.1 days were calculated for TBT and DBT, respectively, whereas apparent t(1/2) values could not be determined for MBT and TPhT (t(1/2)>18 days).

The capacity of activated sludge to biodegrade OT compounds in the absence of supplemental substrate indicated that these compounds can be metabolized as single sources of carbon and energy in activated sludge systems. Excluding TBT, the presence of low concentrations of supplemental substrate did not affect the biodegradation potential of activated sludge.

The acclimatization of biomass on OT compounds enhanced significantly biodegradation, resulting in significant decreases of half-lives of OT compounds. As a result in the presence of acclimatized biomass, half-lives of 1.4, 3.6, 9.8 and 5.0 days were calculated for TBT, DBT, MBT and TPhT, respectively.

Genes Involved in Aniline Degradation by Delftia acidovorans Strain 7N and Its Distribution in the Natural Environment; Urata M et al.; Aniline-degraders were isolated from activated sludge and environmental samples and classified into eight phylogenetic groups. Seven groups were classified into Gram-negative bacteria, such as Acidovorax sp., Acinetobacter sp., Delftia sp., Comamonas sp., and Pseudomonas sp., suggesting the possible dominance of Gram-negative aniline-degraders in the environment. Aniline degradative genes were cloned from D. acidovorans strain 7N, and the nucleotide sequence of the 8,039-bp fragment containing eight open reading frames was determined.

Their deduced amino acid sequences showed homologies to glutamine synthetase (GS)-like protein, glutamine amidotransferase (GA)-like protein, large and small subunits of aniline dioxygenase, reductase, LysR-type regulator, small ferredoxin-like protein, and catechol 2,3-dioxygenase, suggesting a high similarity of this gene cluster to those in P. putida strain UCC22 and Acinetobacter sp. strain YAA. Polymerase chain reaction (PCR) and sequencing analyses of GS-like protein gene segments of other Gram-negative bacteria suggested that Gram-negative bacteria have aniline degradative gene that can be divided into two distinctive groups.

Leucobacter chromiireducens sp. nov, and Leucobacter aridicollis sp. nov., two new species isolated from a chromium contaminated environment; Morais PV et al.; Two strains designated strains L-1T and L-9T were isolated from activated sludge of a treatment plant that receives wastewater from the tannery industry contaminated with chromium.

Phylogenetic analysis showed that the organisms represented two new species of the genus Leucobacter. Strains L-1T and L-9T could be distinguished from the type strain of L. komagatae and from the type strain of "L. albus"by the B-type peptidoglycan composition, fatty acid composition, several phenotypic and physiological characteristics. The major fatty acids of the organisms were iso- and anteiso-branched C15:0 and C17:0, straight-chain C16:0 was also found in relatively high proportions.

The organisms were halotolerant, grew in medium containing 9% NaCl, and all strains, including the type strain of L. komagatae grew in medium containing 5 mM Cr(VI). On the basis of the distinct peptidoglycan composition, 16S ribosomal DNA sequence analysis, percentage of DNA-DNA reassociation values, and phenotypic characteristics we are of the opinion that strain L-1T represents a new species of the genus Leucobacter for which we propose the name Leucobacter chromiireducens and that strain L-9T represents an additional new species of the same genus for which we propose the name Leucobacter aridicollis.

The solids retention time-a suitable design parameter to evaluate the capacity of wastewater treatment plants to remove micropollutants; Clara M et al.; Micropollutants as endocrine disrupting compounds (EDC) or pharmaceuticals are of increased interest in water pollution control. Wastewater treatment plant (WWTP) effluents are relevant point sources for residues of these compounds in the aquatic environment. The solids retention time (SRT) is one important parameter for the design of WWTPs, relating to growth rate of microorganisms and to effluent concentrations.

If a specific substance is degraded in dependency on the SRT, a critical value for the sludge age can be determined. In WWTPs operating SRTs below this critical value, effluent concentrations in the range of influent concentrations or a distribution according to the adsorption equilibrium have to be expected, whereas in WWTPs operating at SRTs higher than the critical value degradation will occur. Critical SRTs were determined for different micropollutants, indicating that the design criteria based on the sludge age allows an estimation of emissions. Different treatment technologies as conventional activated sludge systems and a membrane bioreactor were considered and no significant differences in the treatment efficiency were detected when operated at comparable SRT.

The results of the investigations lead to the conclusion that low effluent concentrations can be achieved in WWTPs operating SRTs higher than 10 days (referred to a temperature of 10 degrees C). This corresponds to the requirements for WWTPs situated in sensitive areas according to the urban wastewater directive of the European Community (91/271/EEC) in moderate climatic zones.Reduction of bromate to bromide coupled to acetate oxidation by anaerobic mixed microbial cultures; van Ginkel CG et al.; Bromate, a weakly mutagenic oxidizing agent, exists in surface waters. The biodegradation of bromate was investigated by assessing the ability of mixed cultures of micro- organisms for utilization of bromate as electron acceptor and acetate as electron donor. Reduction of bromate was only observed at relatively low concentrations (<3.0mM) in the absence of molecular oxygen. Under these conditions bromate was reduced stoichiometrically to bromide.

Unadapted sludge from an activated sludge treatment plant and a digester reduced bromate without lag period at a constant rate. Using an enrichment culture adapted to bromate, it was demonstrated that bromate was a terminal electron acceptor for anaerobic growth. Approximately 50%of the acetate was utilized for growth with bromate by the enrichment culture. A doubling of 20h was estimated from a logarithmic growth curve. Other electron acceptors, like perchlorate, chlorate and nitrate, were not reduced or at negligible rates by bromate-utilizing microorganisms.

Fate of Ah receptor agonists during biological treatment of an industrial sludge containing explosives and pharmaceutical residues; Gustavsson LK et al.; Goal, Scope and Background: Sweden is meeting prohibition for deposition of organic waste from 2005. Since 1 million tons of sludge is produced every year in Sweden and the capacity for incineration does not fill the demands, other methods of sludge management have to be introduced to a higher degree. Two biological treatment alternatives are anaerobic digestion and composting. Different oxygen concentrations result in different microbial degradation pathways and, consequently, in a different quality of the digestion or composting residue, It is therefore necessary to study sludge treatment during different oxygen regimes in order to follow both degradation of compounds and change in toxicity.

In this study, an industrial sludge containing explosives and pharmaceutical residues was treated with anaerobic digestion or composting, and the change in toxicity was studied. Nitroaromatic compounds, which are the main ingredients of both pharmaceutical and explosives, are well known to cause cytotoxicity and genotoxicity.

However, little data are available concerning sludge with nitroaromatics and any associated dioxin-like activity. Therefore, we studied the sludge before and after the treatments in order to detect any changes in levels of Ah receptor (AhR) agonists using two bioassays for dioxin-like compounds. Methods: An industrial sludge was treated with anaerobic digestion or composting in small reactors in a semi-continuous manner.

The same volume as the feeding volume was taken out daily and stored at –20 degrees C. Sample preparation for the bioassays was done by extraction using organic solvents, followed by clean up with silica gel or sulphuric acid, yielding two fractions. The fractions were dissolved in DMSO and tested in the bioassays. The dioxin-like activity was measured using the DR-CALUX assay with transfected H4IIE rat hepatoma pGudluc cells and an EROD induction assay with RTL-W1 rainbow trout liver cells.

Reduction of excess sludge production by 3,3′,4′, 5-tetrachlorosalicylanilide in an activated sludge process; Ye FX et al.; The potential of 3,3′,4′,5- tetrachlorosalicylanilide (TCS) addition to an activated sludge continuous process to reduce excess sludge production by disrupting coupling between anabolism and catabolism was investigated. TCS was chosen as a metabolic uncoupler for continuous test in a lab-scale completely mixed activated sludge process. TCS reduced sludge yield by approximately 30% at a dosage of 40 mg/day. Substrate removal capability was not adversely affected by the presence of TCS, but effluent nitrogen concentration increased during the 60- day continuous operation. Sludge settleability of treated and control samples was qualitatively comparable and not significantly different. Microbial activities in terms of specific oxygen uptake rate were also enhanced, and the microbial population was altered. The results suggest that TCS is an effective chemical uncoupler that reduces sludge yield; process performance was not significantly affected by introduction of the uncoupler.

Stability of the fluorogenic enzyme substrates and pH optima of enzyme activities in different Finnish soils; Niemi RM et al.; Fluorogenic artificial substrates facilitate sensitive enzyme activity measurements for a variety of processes in soil and other environmental samples. It is possible to use in situ pH for measurements on condition that the substrates are chemically stable. We studied the stability of 12 different methyl umbellipherone (MUF) and amino methyl coumarine (AMC) derivatives used as substrates for arylsulphatase, alpha-glucosidase, beta-glucosidase, beta-xylosidase, cellobiosidase, chitinase, phosphomonoesterase (PME), phoshodiesterase (PDE), esterase, lipase and alanine- and leucine aminopeptidases (AP) over the pH range from 4.0 to 8.0 in modified universal buffer (MUB).

Stability of the substrates for lipase (4-MUF-heptanoate) and esterase (4-MUF-acetate) measurements was poor, especially at the higher pH values. Chitinase substrate, 4-MUF-N-acetyl-beta-d-glucosamide, was unstable at high pH values whereas the substrate for PME activity measurement (4-MUF- phosphate) disintegrated at low pH.

The other substrates and MUF and AMC standard solutions were stable over the pH range studied. The optima between pH 4 and 8 of the 11 different enzyme activities were measured in three forest and two agricultural soil samples and in one activated sludge sample. In soil, for alanine and leucine AP the pH optima were usually 7.5 or higher, for arylsulphatase, beta-glucosidase, beta-xylosidase, esterase and PDE between 4 and 5.5, and for cellobiosidase between 4 and 5. alpha-Glucosidase had an optimum below 5.5 but also exhibited high activity at pH 7. Soil-dependent variation in pH optima were observed for chitinase, esterase, PDE and PME. Enzyme activities were also measured in 0.5 M acetate buffer at pH 5.5. This buffer yielded the highest activities in all soil samples for arylsulphatase, PDE and PME.

Persistent organic pollutants (POPs) in the conventional activated sludge treatment process: fate and mass balance; Katsoyiannis A et al.; The fate and the mass balance of persistent organic pollutants (POPs) during the conventional activated sludge treatment process were investigated in the wastewater treatment plant of the city of Thessaloniki, northern

Greece. The POPs of interest were 7 polychlorinated biphenyls and 19 organochlorine pesticides. Target compounds were determined at six different points across the treatment system: the influent, the effluent of the primary sedimentation tank, the effluent of the secondary sedimentation tank, the primary sludge, the activated sludge from the recirculation stream, and the digested/dewatered sludge.

The distribution of POPs between the dissolved and the adsorbed phases of wastewater and sludge was investigated. A good linear relationship between the distribution coefficients, K(d), and the octanol-water partition coefficients, K(ow), of the solutes was observed only in raw wastewater, suggesting that other factors affect the phase distribution of organic compounds in treated wastewater. For all POPs, a significant increase in partitioning with a decreasing solids concentration was observed, revealing an effect from non- settling microparticles remaining in the "dissolved" phase during the separation procedure.

A good linear relationship was also revealed between logK(d) and the dissolved organic carbon (DOC) content of wastewater, suggesting that DOC Favours the advective transport of POPs in the dissolved phase. Almost all POPs showed good mass balance agreements at both the primary and the secondary treatment. The losses observed for some species could be attributed to biodegradation/biotransformation rather than volatilization. The relative distribution between the treated effluent and the waste sludge streams varied largely among different compounds, with p-p'-DDE being highly accumulated in the waste sludge (98%) but almost 60% of alpha-HCH remaining in the treated effluent.

Sludge as source of energy and revenue; Onyeche TI; Sludge is a residue/ product from wastewater treatment plants and contains most of the contaminants released during human activities. Some stringent environmental regulations on sludge treatment and disposal exist in many countries. This has resulted in increasing interest in sludge treatment methods that encourage sludge reduction and improvement in biogas production during anaerobic digestion processes. This work demonstrates the first exploitation of valuable energy from homogenised sludge at technical scale with mass reduction. The optimal combination of sludge homogenisation at relatively low pressures using a modified high-pressure homogeniser led to the success of this unique project. Results showed that about 30% more energy could be obtained from thickened and disrupted sludge than from untreated samples. The energy produced was higher than that invested during disruption and digestion processes. About 23% sludge reduction was also observed with no increase in chemical oxygen demand.

This new process can produce extra energy for local electrification and heating the digester while the sludge reduction provides economic benefits. Concentration of sludge causes reduction in investment cost on digester as well as reduction in operational time for sludge dewatering.

Direct generation of electricity from sludges and other liquid wastes; Dentel SK et al.; Recently it has been shown that electrical energy can be harvested from marine sediments, simply by connection of an electrode (anode) in anaerobic marine sediments to an electrode in the aerobic zone above the sediments. We have now shown that similar applications are available in sludge treatment. Using a reactor with graphite foil electrodes in an aerated aerobic and anaerobic sludge zone, electrical current was generated, and enhanced when an additional organic substrate (acetate) was added. Electron microscopy, x-ray diffraction, and PCR examination of the anode surface showed no surface colonization and no increase in Geobacter relative to a control, indicating that microbial use of the anode as an electron acceptor wa indirect through the use of redox mediators. Given the demonstration of electricity generation from sludge, the potential for similar applications, using other organic waste sources, is evaluated.

Physical characteristics of a waste activated sludge: conditioning responses and correlations with a synthetic surrogate; Dursun D et al.; The efficient and economic management of waste activated sludge (WAS) requires a proper understanding of the sludge's material properties. Though there has been much study of WAS, an adequate linkage between its physical and chemical properties has been elusive. In particular, the conditioning and dewatering of WAS are expensive operations, and the addition of polymer to WAS leads to even more complex material Behaviour that is difficult to optimize. This chapter reports on an extensive characterization of WAS, both with and without polymer conditioning. We combined the classical "jar test" approach with less conventional rheometric and electrokinetic measurements. In addition to the use of sampled WAS, a synthetic surrogate sludge was formulated, attempting to duplicate properties of WAS and allow more extensive characterization of a reproducible surrogate to WAS. Results with both the synthetic surrogate and WAS indicated that the traditional, electrokinetic, and rheological properties were related. However, the dose optima by rheometry were somewhat higher in both cases. Minimization of sludge production and stable operational condition of a submerged membrane activated sludge process; Bhatta CP et al.; Modification of conventional activated sludge process by adding microfiltration type hollow fibre membrane module offers various advantages. The membrane module is immersed directly in the bioreactor to separate the treated water from the sludge and the modified process is called submerged membrane activated sludge process

(SMASP). Bench scale SMASP units were operated at constant flux to study the possibility of minimizing the excess sludge production and stable operation at different SRT and BOD loading.

The long SRT developed high stabilized MLSS concentration in the bioreactor and decreased the sludge loading rate {kg-BOD/kg-MLSS-d}. In SMASP, very low sludge yield coefficients (0.04-0.09 kg-MLSS/kg-BOD) were obtained at long SRT (500 d) and small BOD loading (< or =0.5 kg/(m^3 × d)) as compared to conventional activated sludge process and these coefficients showed considerable minimization in excess activated sludge production. At the same operation conditions, suction pressure was also maintained low (<25 kPa), and then SMASP was operated stably for long time (approximately 200 d).Co-conditioning and dewatering of alum sludge and waste activated sludge; Lai JY et al.; Co-conditioning and dewatering Behaviours of alum sludge and waste activated sludge were investigated. Two different sludges were mixed at various ratios (2:1; 1:1; 1:2; 1:4) for study. Capillary suction time (CST) and specific resistance to filtration (SRF) were utilized to assess sludge dewaterability. Relatively speaking, waste activated sludge, though of higher solid content, was more difficult to be dewatered than alum sludge.

It was found that sludge dewaterability and settlability became better with increasing fraction of alum sludge in the mixed sludge. Dosage required of the cationic polyelectrolyte (KP-201C) for dewatering was reduced as well. It is proposed that alum sludge acts as skeleton builder in the mixed sludge, and renders the mixed sludge more incompressible which is beneficial for sludge dewatering. e, a, g. Implications of the results of the study to the sludge management plan for Taipei City that generates both alum sludge and waste activated sludge at significant amount are also discussed. The current sludge treatment and disposal plan in Metropolitan Taipei could be made more cost- effective.

Ultrasonic sludge treatment for enhanced anaerobic digestion; Hogan F et al.; Ultrasound is the term used to describe sound energy at frequencies above 20 kHz. High-powered ultrasound can be applied to a waste stream via purpose-designed tools in order to induce cavitation. This effect results in the rupture of cellular material and reduction of particle size in the waste stream, making the cells more amenable to downstream processing. sonix is a new technology utilising high-powered, concentrated ultrasound for conditioning sludges prior to further treatment. This chapter presents recent results from a number of demonstration and full-scale plants treating thickened waste activated sludge (TWAS) prior to anaerobic digestion, therefore enhancing the process. The present studies have proved that the use of ultrasound to enhance anaerobic digestion can be achieved at full scale and effectively result in the TWAS (typically difficult to digest) behaving, after sonication, as if it were a "primary"sludge. The technology presents benefits in terms of increased biogas production, better solids reduction, improved dewatering characteristics of the digested sludge mixture and relatively short payback periods of two years or less subject to the site conditions and practices applicable at that time.

Three-dimensional dual-morphotype species modeling of activated sludge flocs; Martins AM et al.; An individual-based model, originally developed for a biofilm system, was adapted to simulate three-dimensional formation of activated sludge flocs. The model was extended to two different bacterial morphotypes (floc-forming and filamentous bacteria), allowing spatial development of the floc according to the bacterial morphology, diffusion, reaction, and growth processes.

The model needed also extension with a process for attachment of individual cells. Despite being in an early stage of development, the model is already a tool that enables us to obtain useful information about the microfloc environment. The model indicates that filamentous bacterial morphology and substrate microgradients are important aspects in the formation of bacterial structures.

In mass transport-limited regimes filamentous bacterial structures prevail, whereas in growth-limited regimes irregular shaped flocs with fingerlike structures are dominant. These modeling results suggest that activated sludge flocs and biofilms might be different manifestations of the same phenomena. The model results support the hypothesis that floc-macrogradients can be the most important parameter for development of bulking sludge.

The model suggests that attachment has a very strong effect on floc structure, leading to enhancement of the effect of substrate microgradients.

Physicochemical properties and stability of activated sludge flocs under temperature upshifts from 30 to 45 degrees C; Morgan-Sagastume F et al.; The impacts of temperature shifts from 30 to 45 degrees C on the structural stability and surface charge of activated sludge flocs were assessed in four sequencing batch reactors (SBRs) treating pulp and paper mill effluent. The improvement in floc stability was tested by sludge magnesium enrichment in one SBR and by operating another reactor at a high sludge retention time (SRT) of 33 days. Floc stability was characterized by dissociation constants with solutions of CaCl(2), KCl, urea, and ethylenediamine tetraacetate (EDTA).

Surface charge was assessed by cationic-anionic titration and metals concentrations were also determined. The temperature shift consistently caused an increase in the negative sludge surface charge from approximately –0.180 to –0.300 meq/g MLSS.

Magnesium enrichment and a high SRT of 33 days promoted less negatively charged sludge, dampened the increase in negative sludge surface charge, and yielded structurally stronger flocs; however, sludge deflocculation still occurred. Manganese and iron appeared to be released by sludge under the temperature shift.

It was concluded that the temperature shift deteriorates the flocculating physicochemical properties of the sludge and that better floc stability achieved by magnesium enrichment and a high SRT is not enough to stop deflocculation. Further research is required to clarify the origin of the increase in negative sludge surface charge, the role of metals, and the governing factors in sludge deflocculation under such temperature shifts.

Optimized sequence retrieval from single bands of temperature gradient gel electrophoresis profiles of the amplified 16S rDNA fragments from an activated sludge system; Zhang X et al.; Sequence retrieval from single bands of polymerase chain reaction (PCR)-denaturing gel electrophoresis (DGE) profiles is an important but often difficult step for molecular diversity analysis of complex microbial communities such as activated sludge systems.

We Analysed the temperature gradient gel electrophoresis (TGGE) profiles of PCR-amplified 16S rDNA fragments from an activated sludge sample of a coking wastewater treatment plant. Single bands were excised, and a clone library was constructed for each.

Sequence heterogeneity in each single band was found to be significantly overestimated due to single-stranded DNA (ssDNA) contamination formed during the PCR amplification, since only 10-60% of library clones of each single TGGE band had identical migration Behaviour compared with the parent band.

Three methods, digestion with mung bean nuclease, optimization of PCR amplification, and purification via denatured polyacrylamide gel electrophoresis (d-PAGE), were compared for their ability to minimize ssDNA contamination, with the last one being the most efficient. After using d-PAGE to minimize ssDNA to a nearly nondetectable level, 70-100% of library clones for each single TGGE band had identical migration compared with the parent band.

Several sequences were found in each of six single bands, and this co-migration could be predicted with the Poland software. The predominant bacteria of the activated sludge were assessed via a combination of sequence retrieval from each single TGGE band and band intensity analysis. Only beta and alpha subclasses of the Proteobacteria were detected, 93.8% and 6.2%, respectively. Our work suggests that prior to constructing a clone library to retrieve the actual sequence diversity of a single DGE band, it is advisable to minimize ssDNA contamination to a nondetectable level.

High rate treatment by aerobic upflow sludge blanket (AUSB) with external oxygenation; Sharma KR et al.; A 3-year study was conducted using an aerobic upflow sludge blanket (AUSB) reactor to achieve high-rate biological treatment through maintenance of a high biomass concentration (7-10 g VSS/ L) together with a high oxygen flux. The AUSB reactor was not aerated directly; instead, oxygenation was provided in an external chamber with flow recirculation. The oxygenation was provided at four different pressurizations: 0, 15, 20 and 25 psig. The AUSB reactor was also pressurized to avoid the flotation of biomass. The flow recirculation rate was varied from 400% to 1,500% in order to ensure adequate oxygen supply. It was found that the AUSB system was able to handle a volumetric loading of as high as 10 kg COD/m3- day with a removal efficiency of 92%. Despite a high upflow velocity through AUSB, the effluent suspended solid concentration was mostly below 60 mg/L. The active fraction of biomass in the AUSB sludge was about 3 times higher than that of the regular activated sludge. This was indicated by a very high specific oxygen uptake rate (SOUR), up to 180-250 mg O2/g VSS-hr. The sludge yield in the entire system was only 0.09 to 0.13 g VSS/g COD removed. This was mainly caused by additional auto-oxidation of biosolids in the oxygenation chamber due to flow recirculation.

Effect of chlorination bulking control on water quality and phosphate release/uptake in an anaerobic-oxic activated sludge system; Chang WC et al.; This study evaluates the effect of chlorination bulking control on water quality and phosphate release/uptake in an anaerobic-oxic activated sludge system. A series of batch experiments with different specific NaOCl mass dose were conducted to determine the sludge settling properties, supernatant water quality and phosphate metabolism Behaviour of filamentous bulking sludge.

The harvested sludge was from a continuous-flow anaerobic-oxic (A/O) activated sludge pilot-plant, *i.e.*, enhanced biological phosphorus removal (EBPR) system, operated with 15 days of sludge retention time. The filamentous bacteria in the A/O pilot plant were identified to be Thiothrix according to Eikelboom's classification techniques, which was in accordance with the high influent sulfate concentration of this study (50 mg/L sulfate). Increasing NaOCI concentration, as revealed by experimental results, obviously decreased the sludge settling properties (SVI values and zone settling velocities) and meanwhile significantly reduced supernatant water quality (COD, SS, TP) mainly due to higher suspended solids caused by floc disruption. Moreover, the nine-hour batch experiments indicated that high NaOCI dosage (40 mg/ gMLSS) completely deteriorated phosphate metabolism of EBPR sludge. Such a high dosage of chlorination further confirmed overdosing through disappearance of intracellular PHB and death of protozoa by microscopic investigation.

Still, phosphate release/uptake Behaviour of EBPR sludge properly functions at low NaOCl dosage (5 mg/g MLSS). Besides, phosphate metabolism worsens rapidly before the SVI value reaches its lowest level. These findings imply that determining NaOCI requirement with merely SVI values can readily result in chlorination overdosing. Proper NaOCI dosage requires a delicately balanced consideration between sludge settling improvement, water quality demand and phosphate metabolism. Batch test of phosphate release/uptake is apparently a prerequisite to conclude an appropriate NaOCl dosage for bulking control.

Microbial degradation of estrogens using activated sludge and night soil- composting microorganisms; Shi JH et al.; In order to investigate the potential for microbial degradation of estrogens, and the products formed, activated sludge collected from Korea (ASK) and night soil-composting microorganisms (NSCM) were used to degrade estrogens. Results showed that both ASK and NSCM degraded almost 100% of the natural estrogens estrone (E1), 17beta- estradiol (E2), and estriol (E3) from initial concentrations of 20-25 mg/L, while synthetic estrogen, ethynylestradiol (EE2), was not degraded. Analysis of degradation products of E2 by using HPLC-ECD and a consecutive first-order reaction calculation confirmed that E2 was sequentially degraded to E1, which was further degraded to other unknown compounds by ASK and NSCM.

We then used the yeast two-hybrid assay to show that the unknown degradation products did not appear to possess estrogenic activity when E1, E2 or E3 were degraded to below the detection limit after 14 days of incubation, indicating that ASK and NSCM not only degrade natural estrogens, but also remove their estrogenic activities. Effects of phosphorus limitation and temperature on PHA production in activated sludge; Chinwetkitvanich S et al.; The study was designed to investigate the effects of temperature and phosphorus limitation on polyhydroxyalkanoate (PHA) production and storage by activated sludge biomass. The two-stage operation approach, *i.e.* a growth phase followed by a nutrient limitation phase, was applied to induce PHA accumulation.

The pre-selected temperatures of 10, 20 and 30 degrees C were investigated under phosphorus limitation conditions using three four-litre fully aerobic SBR systems operated at an SRT of 10 days with cycle time and HRT of 6 and 10 hours. PHA production was greater in the 10 degrees C system than in the 20 degrees C and 30 degrees C systems but there was little difference between the two higher temperatures. The maximum PHA fractions of the sludge were 52, 45 and 47%TSS for the three temperatures from low to high, and the maximum PHA concentrations in the mixed liquors were 1,491, 1,294 and 1,260 mg/l, respectively.

However, it was observed that very low values of PHA yield per unit COD consumed were obtained, *i.e.*, 0.05, 0.03 and 0.04 mgPHA/mgCODu, for the 10, 20 and 30 degrees C reactors, respectively. This was because all three systems required several days to reach maximum PHA accumulation in their mixed liquor biomasses. It is probable the bacteria still had some stored poly- P in their cells upon initiation of the phosphorus limited influent, and PHA accumulation was delayed until the stored phosphorus was depleted. Also, PHA productivity was reduced by the large amounts of biomass lost from the systems because of sludge bulking.

Sludge thickening performance of mesh filtration process; Park MS et al.; Small-scale wastewater treatment facilities play an important role in improving the aquatic environment in many countries. Although sludge treatment is essential for overall wastewater treatment, it is difficult for small-scale facilities to use mechanical equipment or other facilities. As the first step of the sludge treatment, it is important to develop a convenient sludge thickening process for small-scale facilities.

In this work, we examined the sludge thickening performance of a mesh filtration system: the mesh opening sizes of 100-500 microm, and the sludge (3,000-9,000 mg-SS/L) was obtained from a domestic wastewater treatment facility. The filtration was carried out only under the hydraulic pressure between the water level and the effluent port connected to the mesh filter module. The sludge reduction rates were in the range of 85-95% for 6-7 h; the initial filtration rate was very high, but the rate decreased with a decrease in hydraulic pressure due to the reduction of the water level in the vessel. In addition, the effluents (passed through the mesh) contained very low SS and could be directly discharged into the environment.

Biological treatability of raw and ozonated penicillin formulation effluent; Cokgor EU et al.; In the present study, oxidative pre-treatment of pharmaceutical wastewater originating from the formulation of the penicillin Sultamycillin Tosylate Diydrate via ozonation at varying pH and ozone feed rates was investigated. Biological treatability studies were performed with a synthetic wastewater alone and supplemented with raw and ozonated penicillin formulation effluents.

The highest COD (34%) and TOC (24%) removal efficiencies were obtained at pH 11.0, whereas the BOD5 value increased from 16 mg l(–1) to 128 mg l(–1) after 40 min of ozonation, corresponding to an applied ozone dose of 1670 mg l(-1) and 33% relative ozone absorption. The studies showed that no degradation of raw penicillin fraction (30% of total COD) occurred, and degradation of the synthetic wastewater being completely treatable without penicillin addition, was inhibited by 7%. Upon 40 min ozonation, the synthetic wastewater could be completely oxidized and at the same time 35% of ozonated penicillin wastewater removal was obtained.

Respirometric studies were conducted in parallel and produced results indicating a 22% decrease in the total oxygen consumption rate established for raw penicillin formulation effluent compared to the results obtained from the aerobic batch reactor. No inhibition of the synthetic fraction was observed for the 40 min-ozonated penicillin formulation effluent, biodegradability of the 60 min-ozonated penicillin effluent decreased possibly due to recalcitrant oxidation product accumulation.

The modeling study provided experimental support and information on inhibition kinetics in activated sludge model no. 3 (ASM3) by means of respirometric tests for the first time. Reduction of coagulant amount added to activated sludge for phosphorus removal; Nakajima J et al.; Adding coagulant to the activated sludge process is effective in maintaining the stability of phosphorus removal. However, the precise mechanisms of the reaction and Behaviour of coagulants and phosphorus are not well known. By introducing a new phosphorus removal model (PRM), the Behaviour of coagulant and phosphorus in the process could be described. The experimental data of the effluent phosphorus concentration and Fe content in the activated sludge agreed with the values calculated by PRM. The amount of coagulant addition to the activated sludge process for phosphorus removal is reduced with the enhanced biological phosphorus removal process. It is suggested that the amount of reduction is determined by using PRM.

Monitoring activated sludge settling properties using image analysis; Jenne R et al.; The goal of this study is to develop a monitoring system for activated sludge properties, as this is an essential tool in the battle against filamentous bulking. A fully automatic image analysis procedure for recognising and characterising flocs and filaments in activated sludge images has been optimised and subsequently used to monitor activated sludge properties in a lab-scale installation. The results of two experiments indicate that the image information correlates well with the Sludge Volume Index. It is shown that, at the onset of filamentous bulking, there is an increase in total filament length on the one hand, and a change in floc shape on the other hand.

Performances of three R-AN-D-N wastewater treatment plants in the Czech Republic; Bejvl Z et al.; The evolution of Czech standards requires higher efficiency of nutrient removal from municipal wastewaters. At the beginning of the last decade of 20th century, a new activated sludge configuration called R-AN-D-N process has been described, successfully tested and now largely used at several wastewater treatment plants (WWTP) in the Czech republic.

The main feature of the R-AN-D-N process is the introduction of a regeneration zone in sludge recycle, which enables to increase sludge age in the system without any substantial increase in WWTP volume. Performances of three Czech large WWTP with R-AN-D-N configuration have been monitored and compared within a period of one and a half years.

The results confirmed excellent nutrient removal efficiency for wastewaters with different proportion between sewage and industrial effluents. Two of three monitored WWTP received wastewaters from breweries (Budweiser and Pilsner Urquell). The settleability of activated sludge from all three WWTP was correct, with SVI values usually ranging from 50 to 150 ml/g. Monitoring of sludge composition indicated proliferation of several filamentous bacteria, particularly types 0581, 0092 and M. parvicella. No severe bulking events were observed. Finally, the operational costs expressed in CZK (Czech crown: 1 CZK = {see symbol in text}0.0322) per cubic metre of treated sewage or per capita amounts respectively from 2.24 to 6.52, and from 285 to 342.

North American performance experience with anoxic and anaerobic selectors for activated sludge bulking control; Parker D et al.; Obtaining reliably high compaction characteristics and low SVI values has enormous economic consequences on activated sludge plant design and operation. Now that use of selectors has become more common for achieving low SVI values, the assessment of the full-scale performance characteristics of activated sludge plants incorporating selectors has high value for the design and operation of future plants.

Treatment plant effluent quality is seldom Analysed solely on an average basis and the impact peak events have on effluent quality must be considered in design and operation. Descriptive statistics should also be used to assess selector performance, so that the influence of infrequently occurring values on plant design and operation can be assessed. The performance of 21 activated sludge plants incorporating selectors was evaluated. All the selector plants for which post installation data were available showed that the operating SVI values were significantly improved. One plant studied in detail showed that final effluent quality was significantly better after the selector installation. As a group, activated sludge plants with anaerobic selectors outperform those with anoxic selectors. Dissolved oxygen control is just as important in activated sludge plants with selectors as in conventional activated sludge plants.

Influence of sludge properties and hydraulic loading on the performance of secondary settling tanks--full-scale operational results; Vestner RJ et al.; Full-scale investigations at a WWTP with a two-stage secondary settling tank process revealed relationships between significant operating parameters and performance in terms of effluent suspended solids concentration. Besides common parameters (*e.g.* surface overflow rate and sludge volume loading rate) feed SS concentration and flocculation time must be considered. Concentration of the return activated sludge may help to estimate the performance of existing secondary settling tanks.

Modeling of a large-scale wastewater treatment plant for efficient operation; Gokcay CF et al.; Environmental legislations in the Western world impose stringent effluent quality standards for ultimate protection of the environment. This is also observed in Turkey. The current paper presents efforts made to simulate an existing 0.77 million m3/day conventional activated sludge plant located at Ankara, AWTP. The ASM1 model was used for simulation in this study. The model contains numerous stoichiometric and kinetic parameters, some of which need to be determined on case by case bases. The easily degradable COD (S(S)) was determined by two methods, physical-chemical and respirometric methods, namely. The latter method was deemed unreliable and rejected in the further study. Dynamic simulation with SSSP Programmeme predicted effluent COD and MLSS values successfully while overestimating OUR. A complete fit could only be obtained by introducing a dimensionless correction factor (etaO2 = 0.58) to the oxygen term in ASM1.

A control strategy for reducing aeration costs during low loading periods; Sahlmann C et al.; The efficiency of the aeration system in a full-scale activated sludge basin with 3 separately controlled aeration zones was improved for the low loading period in summer. The air flow rate to each aeration zone is currently regulated to hold a preset dissolved oxygen concentration (DO). Four different DO setpoint combinations were tested, each one for a one week period, using dynamic off-gas testing to measure the standardised oxygen transfer efficiency (alphaSOTE). As the DO setpoints were lowered, the total air flow rate to the basin decreased initially.

A low DO in the first zones slowed biomass activity and pushed the load towards the end of the aeration basin. The relationship between alphaSOTE and the specific diffuser flow rate qD is different for each zone. In Zone 1 there was a strong decrease in alphaSOTE as qD increased, while Zones 2 and 3 were fairly independent of qD, Zone 2 at a higher level than Zone 3. Aeration costs were reduced by 15% for the most efficient combination. To achieve even more savings, a control strategy adjusting oxygen transfer rates over the aeration basin to the necessary oxygen transfer rates is suggested. It is based on changing the DO setpoints to reach the lowest total air flow rate while meeting the effluent requirements.

Nutrient removal process development and full scale implementation at the 4 million p.e. main treatment plant of Vienna, Austria; Kroiss H et al.; The Main Treatment Plant of Vienna is in extension for 4 million p.e. and very stringent nutrient removal requirements. The existing high rate BOD removal activated sludge plant (in operation since 1980) is extended by a second stage activated sludge plant and a newly developed flow scheme for nitrogen removal optimisation adaptable to the temperature variations over the year. For this plant pilot investigations have been performed for the development of a specific mathematical model (ASMV) and a specific aeration control strategy.

The civil work of the extension is already finished and the installation of the equipment has started. Operation should start in 2004. The whole project will cost about {see symbol in text 264 million of which about one half is for civil work. The effluent standards correspond to the requirements for sensitive areas in EU Directive for Municipal Waste Water. The raw primary and excess sludge are incinerated after thickening and dewatering. This chapter tries to condense the already existing literature with the construction progress and the cost situation. Isolation and characterization of sulphur-oxidizing Thiomonas sp. and its potential application in biological deodorization; Chen XG et al.; AIMS: To isolate and characterize a sulphur-oxidizing bacterial strain from activated sludge and to evaluate its potential application in biological deodorization. Methods and Results: A dominant sulphur-oxidizing bacterial strain, designated as strain SS, was isolated from an enrichment culture using thiosulphate as a sole energy source and CO_2 as a sole carbon source.

The cells of this organism were aerobic, rod-shaped, Gram-negative and motile. Strain SS could grow autotrop-hically, heterotrophically as well as mixotrophically. Autotrophic growth was observed at pH values ranging from to 9.0. Phylogenetic analyses revealed that strain SS belonged to Group 1 of the genus Thiomonas, closely related to Thiomonas perometabolis and Thiomonas intermedia. The thiosulphate oxidation rates of strain SS at different pH values were evaluated in terms of oxygen uptake using a Micro-Oxymax respirometer. The results showed that the maximum oxidation rate of 5.65 mg l(−1) h(−1) occurred at 56 h of growth and pH 6.0. Continuous H2S removal study demonstrated that strain SS could remove more than 99% of H2S when the inlet concentration was below 58.6 ppm.

Further increase of the inlet concentration to 118 ppm gave rise to a decline in the removal efficiency to ca 90%. Conclusions: The strong acidification of the culture medium during the later period could result in the deterioration of the growth activity and the metabolism activity of strain SS. In practical application, the problems caused by the end-product inhibition and the acidification can be alleviated by periodical replacement of culture medium with fresh medium. d, b. Given the physiological flexibility and the ability to remove H2S rapidly and efficiently, strain SS could be a good 'deodorizing' candidate. Significance and impact of the study: This is the first time that Thiomonas species has been reported for biological deodorization application. Start up of deammonification process in one single SBR system; Li X et al.; A process for autotrophic nitrogen removal named aerobic/anoxic deammonification wherein NH_4^+ is oxidized by nearly 50% to NO_2^- andsubsequently the ammonia is converted together with the nitrite to molecular nitrogen (N2 gas), has come to full-scale application within the last few years.In this research, sludge from a biological rotation disk located at a landfill leachate plant at Mechernich, Germany, which is capable of performing the deammonification process, was used as seed sludge for acclimating deammonification activities in Labouratory scale batch-reactors.

In parallel, the same tests were performed with normal activated sludge. Research results indicated that deamm-onification activities could be obtained from the seeded reactor and also, with limited performance, from normal activated sludge in a single SBR system after several months acclimation. It was also seen that oxygen is an important factor that influences the deammonification from both the acclimatization process and process running.

Further results were approved that report an impact of nitrite as a process intermediate on the closely related process of anaerobic ammonia oxidation ("Anammox"). However, limiting concentrations on a bacteria population performing deammonification were found to be different to those reported for a pure Anammox-culture. Also the influence of another intermediate, hydrazine, was tested for speeding up the acclimating process by inducing the deammonification activities and recovering the activities of deammonification from nitrite inhibition. ❐

3

Energy and Environment Perspectives

BIOGAS TECHNOLOGY DISSEMINATION

Despite numerous benefits derived from biogas technology is yet to develop a major programmes that will promote the dissemination of biogas plants on a larger scale. This chapter reviews biogas installations and investigates challenges facing the design, construction, and operation of biogas plants. It further captures the current status and functions of biogas plants as well as the impact of these plants on the people who use them. The study was done by surveying fifty (50) biogas installations, and conducting interviews with both plant users and service providers.

From the survey, twenty-nine (58%) installations were institutional, fourteen (28%) were household units, and the remaining seven (14%) were community plants. Fixed-dome and water-jacket floating-drum digesters represented 82% and 8% of installations surveyed, respectively. It was revealed that sanitation was the main motivational reason for people using biogas plants. Of the 50 plants, 22 (44%) were functioning satisfactorily, 10 (20%) were functioning partially, 14 (28%) were not functioning, 2 (4%) were abandoned, and the remaining 2 (4%) were under construction.

Reasons for non-functioning include non-availability of dung, breakdown of balloon gasholders, absence of maintenance services, lack of operational knowledge, and gas leakages and bad odour in toilet chambers of biolatrines. This chapter recommends the development of a national biogas programmes focussing on three major areas–sanitation, energy, and agricultural fertilizer production; it further supports the development of standardized digester models.

The founding of a national body or the establishment of a dedicated unit within an existing organization with the sole aim of coordinating and managing biogas dissemination is proposed. Active harnessing and development of renewable energy (RE) sources began in the mid 1980s. However, after more than 20 years in the recognition of RE potential in contributing substantially to the energy mix of the developing country still lacks the capacity to developing her own RE sources.

In 1997, the Energy Commission (EC) was founded, among other functions, to develop, regulate, and manage RE. In 2006, the EC developed the Strategic National Energy Plant (SNEP)–a policy document that defined the role of various energy sources, setting target for each within a twenty year span. From SNEP, biomass based energy, apart from the direct use as woodfuel (firewood and charcoal), has been exploited to a very limited extent. Woodfuel represent the traditional energy source and accounts for 60 per cent of total energy used.

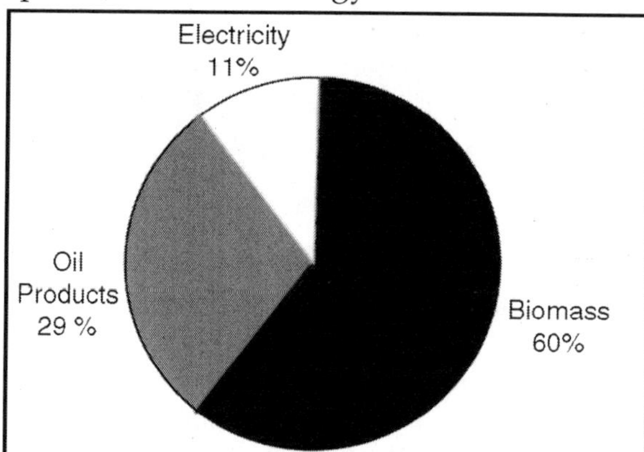

Fig. 3.1: Energy Share of Fuels.

Biofuels (biogas, biodiesel, and bioethanol) have not been adequately developed to play a major role in the energy mix. For instance, in SNEP, the potential contribution of biogas technology towards the growth of the energy sector was not captured compared to other renewable energy options such as wind and solar. In order to realise a reduction in the share of woodfuel in the national energy mix from 60% in 2006 to 40% in 2020 as stipulated in SNEP, there is the need to promote research and development in other renewable energy options including biogas technology.

Biogas (anaerobic fermentation) technology is noted for improving sanitation, generating clean energy, and producing rich organic fertilizer. In China, India, and Nepal, household and institutional biodigesters have gained widespread acceptance. Since 2001, China has disseminated over 2 million household digesters annually; in addition, the Chinese government has supported over 200 large and medium livestock farms to own large and advanced biogas units.

From 2001 to 2007, over 18 million households adopted the technology leading to the production of over 7 billion m^3 of biogas; moreover, 87 million tonnes of animal waste were treated by 3,556 biogas plants and more than 300 Clean Development Mechanism (CDM) projects involving biogas power generation, with a total capacity of 1 GW and an annual emission reduction of over 20 Mt of CO_2, were also developed. In India, over 3 million domestic digesters and 3000 community and institutional plants were constructed by the end of 2002, and since 2005, more than 100,000 biodigesters have been disseminated annually.

Other successful biogas promoting Asian countries include Nepal, Vietnam, and Thailand. In Africa, biogas technology dissemination has been relatively unsuccessful. Njoroge attributes the non-progressiveness of most biogas programmes to failure of African governments to support biogas technology through a focused energy policy, poor design and construction of digesters, wrong operation and lack of maintenance by users, poor dissemination strategies, lack of project monitoring and followups by promoters, and poor ownership responsibility by users. Despite the relative stagnation of biogas programmes in Africa, the future prospects are encouraging.

Aside energy (cooking and lightning, fuel replacement, shaft power), several biogas plants in recent years have been constructed as environmental pollution abatement system in several countries including Ghana, Kenya, Tanzania, Rwanda, Burundi, and South Africa. Between 4000–5000 digesters is estimated to have been built in Tanzani, while Kenya is said to have disseminated about 2000 digesters as at October 2007.

BIOGAS TECHNOLOGY DISSEMINATION

Before 1990

Interest in biogas technology is began in the late 1960s but it was not until the middle 1980s did biogas technology receive the needed attention from government. Dissemination programmes before the mid 1980s focused on the provision of energy for domestic cooking. Most plants, however, collapsed shortly after duration of project due to immature technologies and poor dissemination strategies. In order to resuscitate the technology, a cooperative agreement between Ghana and China led to the construction of a 10 m^3 plant at the Bank of Ghana (BoG) cattle ranch in the Shai Hills and the start of the Appolonia Household Biogas Programmes in 1986.

The Appolonia Household Programmes focused on energy for cooking for cattle-owning households at Appolonia, a rural community in the Greater Accra region. A total of nineteen fixed-dome digesters comprising six 15 m^3 and two 30 m^3 Deenbandhu digesters, and eight 10 m^3 and three 25 m^3 Chinese dome digesters were constructed by engineers from the Ministry of Energy (MoE) and the Institute of Industrial Research (IIR). This was followed by the construction of two household demonstration plants at Jisonayilli and Kurugu, all in the Northern Region in 1987, under the aegis of the United Nations Children Fund (UNICEF). Apart from the Ministry of Energy, biodigesters were also been promoted by Dr. Elias Aklaku, an engineer and a senior lecturer at the Agricultural Engineering Department of KNUST, mostly with support from the German Agency for Technical Cooperation (GTZ).

From 1990 to 1999

In June, 1992, the Ministry of Energy commissioned the first large scale community-based biogas plant in Appolonia. The Appolonia Integrated Rural Energy Project was aimed at providing street lighting and electricity for small load appliances for all the households in the community. Cow dung and human excreta were used to feed the digesters, and the gas produced was used to run a 12.5 kVA generator which provided power for street and home lighting, while the bio-slurry was used for agriculture.

The project experienced several setbacks and did not performed satisfactorily as planned due to multiplicity of factors including feedstock availability problems, distance of kraals (1/2 km) from the community, maintenance problems, and uncooperative attitude of some of the inhabitants. Problems also arose in the utilization of the digested slurry as farm manure. In the initial stages, the liquid organic fertilizer from the plant was successfully used on farms even though farmers complained of intense labour involved in carrying liquid fertilizer from the plant site to their farms.

Another major problem was the drudgery involved in collecting dung from kraals situated hundreds of meters away from the community. Furthermore, Fulani herdsmen prevented women from collecting dung from the kraals on the basis that women could cause pregnant cows to give premature births. Apart from the Ministry of Energy, the Catholic Secretariat and GTZ have been involved in biogas promotion. The Secretariat financed the construction of biogas plants at the Catholic Mission at Kaleo in the Upper West Region; and in three Catholic hospitals–Holy Family and St.

Dominic hospitals in the Eastern Region, and Battor hospital in the Volta Region–between 1994 and 1995. Some biogas projects financed by GTZ has been disseminated to treat slaughterhouse waste at Ejura and KNUST in the Ashanti Region. From 1993 and beyond, direct involvement of the MoE in biogas dissemination slumped mainly due to lack of donor support and unfulfilled expectations of the Appolonia projects. Attempts were made to rekindle the involvement of government in 1996 when the MoE financed a study aimed at assessing biogas resources in Greater Accra, Volta, and the three Northern Regions.

This study was intended to be the first step in planning and developing a national biogas programmes. Ampofo estimated a potential of 88,144 m^3 of biogas a day which could generate about 193 GWh of energy annually. Ampofo did not estimate the quantum of liquid fertilizer that could be generated but Bensah and Brew-Hammond have shown that an amount of 360,000 tonnes of liquid organic fertilizer could be produced yearly, which would be capable of fertilizing about 70,000 hectares of irrigated farmland or 140,000 hectares of dry farmland. Intriguingly, after more than a decade since the study was completed and the report submitted to the MoE, the Ministry has shown no interest in developing a national biogas programmes. Furthermore, the Institute for Industrial Research, a parastatal organization

From 2000 to Date

According to Edjekumhene *et al*, barriers that have plunged biogas dissemination include unfavourable policies, non-availability of feed materials, poor financing arrangements, problems with social acceptance, absence of market, and lack of information. Following the low involvement of biogas projects by government, a number of private biogas companies have marketed the technology on purely business grounds, and mainly based on the ability of biogas plants to improve sanitation. The two leading companies–Biogas Technology West Africa Limited and Beta Civil Engineering Limited–have been constructing brick-based fixed-dome and concrete-based Puxin digesters, respectively. According to SNV, the potential to realise the dissemination of about 270 thousand domestic biogas plants. This is contained in the Biogas for Better Life (B4BL) initiative for Africa–an ambitious programmes conceived by developing partners of Africa and launched in Nairobi in May 2007.

The major aim is to combat poverty by providing over two million households in Africa with biogas plants, with collateral benefits of improved family health through reduction of indoor pollution and drudgery involved in firewood collection and usage. Other objectives of the initiative include the establishment of more than 800 private biogas companies and 200 biogas appliance manufacturing workshops and the construction of one million biotoilets.

In a study to assess domestic biogas potential from cow dung in the three Northern and Ashanti Regions, KITE estimated a technical potential of 80,000 household biogas installations and a market potential (estimated based on the ability and willingness of users to pay) of about 8,000 (8% of 80,000) plants. According to KITE, households have not shown much interest in using biogas in their kitchen or the digested slurry in their farms; thus, biogas technology has been disseminated mostly in institutions such as schools, hospitals, prisons, and slaughterhouses.

Current State of Biogas Technology

A survey of 50 biogas plants was conducted in order to ascertain the true state of biogas technology. Field visits to biogas installations were conducted between June, 2008 and February, 2009. The sample size (50 plants) was determined from the population (100 known biogas plants as captured in a survey by KITE) using stratified and convenience sampling techniques.

The sample size was selected from the strata using convenience sampling. The survey technique used included direct observation of the various components of biogas installations, and structured and unstructured open- ended interviews with both users and experts. Major challenges encountered during the field visits included inadequate funding, difficulty in locating biogas plants in cities, towns and villages where they have been disseminated, inability of service providers to provide full records of the location and number of plants constructed, and the absence of a national body that keeps track of developments in the industry.

All the aforementioned challenges affected the smooth gathering of data and information; nonetheless, the depth of data gathered was enough to make generalizations on biogas technology dissemination. Figure shows the number of installations built over the last three decades as captured by the survey.

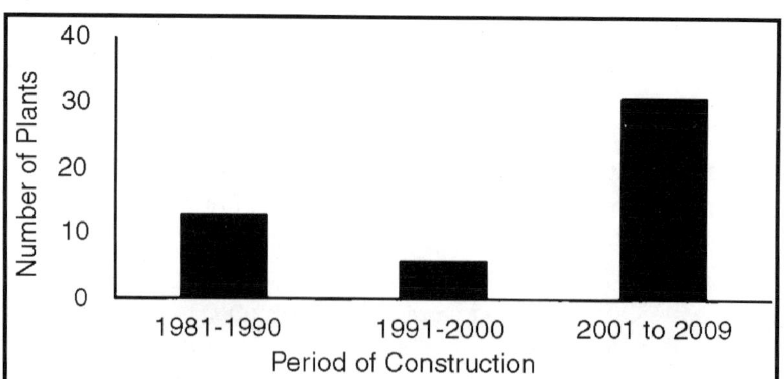

Fig. 3.2: Period of Construction of Surveyed Plants

The 1981-1990 period saw the implementation of the domestic biogas programmes at Appolonia and Okushibli–two major cattle-rearing communities in Greater Accra Region. In 1992, the Appolonia Electricity programmes was commissioned; in addition, biogas plants were built in three hospitals at Battor, Nkawkaw, and Akwatia. The relatively large number of plants disseminated from 2001 to 2009 can be ascribed to the founding of BTWAL, the largest biogas company, by John Idan in 2000. Some of the biogas installations disseminated by BTWAL can be found at the Golden Jubilee House (Presidential Palace) in Accra, Central University College–Miotso campus, and Tamale Teaching hospital. Another major factor was the decision by Beta Civil Engineering Limited–a construction company–to disseminate biogas plants; the company has constructed over 40 Puxin digesters and a few in Nigeria since 2006.

Apart from the aforementioned companies, the other biogas service providers have not been very active; for instance, IIR (Institute of Industrial Research) has disseminated a few biotoilets at various locations between 2004 and 2006, in addition to a 200 m^3 plant at Ankarful Prison in the Central Region. Our of 50 installations studied, 22 were in good condition, 10 were functioning even though some defects (including deteriorated gasholders, gas pipelines and appliance) were observed, and 14 were broken-down. This observation is shown in Figure.

The oldest operating plants were found at the St. Dominic Catholic hospital which has been functioning uninterrupted for 15 years despite intermittent problems with the gas delivery systems.

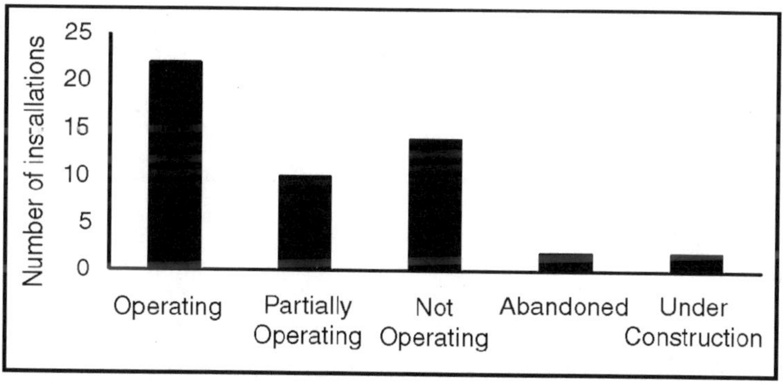

Fig. 3.3: Functional status of surveyed installations

More than 50% of surveyed installations were institutional plants while educational and health institutions accounted for 55% of the 29 institutional units. It was also observed that majority of plants (76%) had been constructed mainly in the cities for the treatment of human excrement from flushing toilets. At Obuasi, Anglogold Ashanti Limited has adopted biosanitation technology in treating sewage from her estates and other facilities including the company's hospital.

With the exception of the plants constructed under the Appolonia Electrification Programmes, all community plants surveyed were biolatrines (pit latrines attached to biogas plants).

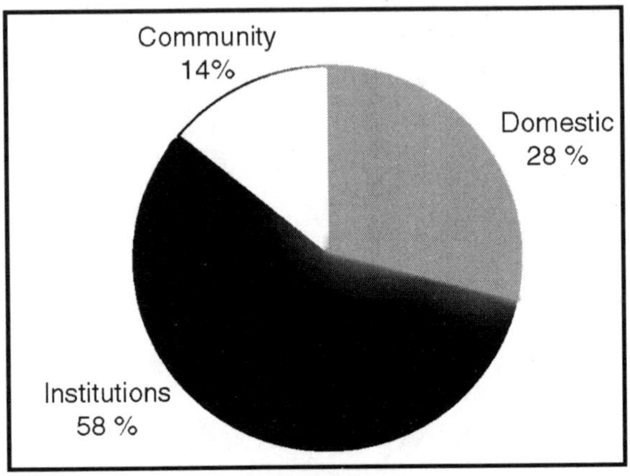

Fig. 3.4: Surveyed Installations Grouped into Institutional, Community, and Domestic Plants

The domestic installations surveyed include six plants built in 1987 at Appolonia and five plants constructed at Okushibli in 1990; most of these installations broke down between 1996 and 2001 due to several challenges including non-availability of cow dung, availability of woodfuel, deterioration of gas stoves, and lack of follow-up services by biogas service providers. After the Appolonia Electrification programmes in 1992, no community installation digesting cow dung has been constructed due to the high cost of biodigesters, reduction of financial support from GTZ, and the absence of specific programmes designed to promote community digesters in cattle-rearing terrains.

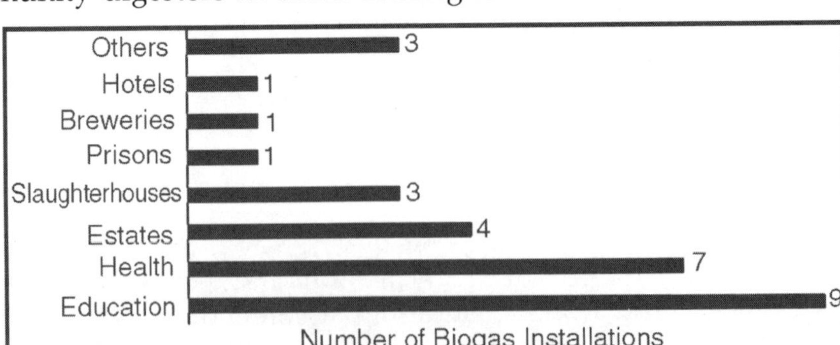

Fig. 3.5: Institutions using Biogas Systems

MOTIVATION FOR USING BIOGAS PLANTS

As part of the survey, biogas users were asked to state the main motivational factor that influenced them to adopt biogas technology. The response is shown in Figure. It is obvious that most users of biogas units patronize the technology solely on the ability of biodigesters to treat nightsoil, thereby replacing septic tanks.

Biogas service providers have achieved success in promoting the technology as the most efficient and cost effective way of treating sewage despite the high cost of biodigesters. There are a number of sewage rehabilitation projects where old sewage handling systems such as septic tanks and Kumasi Ventilated Improved Pits (KVIPs) are gradually making way for biosanitation interventions.

Apart from improving sanitation, biogas plants are also known for generating energy and organic fertilizer. It appears that the ability of biogas plants to produce organic fertilizer for agriculture has not been given priority by service providers. Most users of biogas systems are content with the performance of the plant as long as the plant treats their sewage without any problem. The low level of interest to disseminate biogas plants focusing on energy and agriculture can be attributed to lack of a concerted programmes targeting the dissemination of biogas plants among livestock farmers, majority of them living in rural communities and with limited/no access to agricultural inputs and modern energy services. These are people who will value the products of the biogas plant and therefore make use of both the gas and the fertilizer.

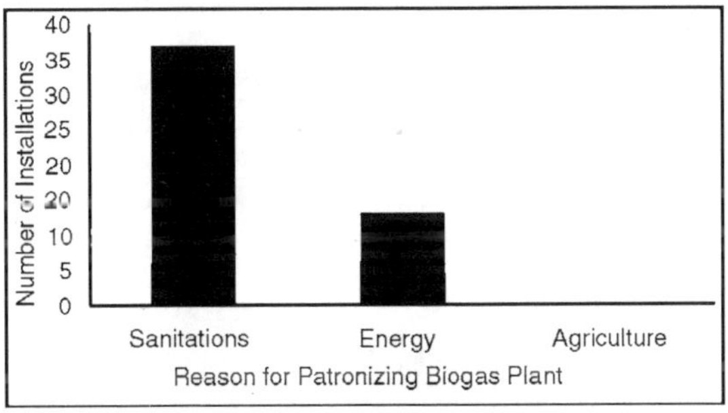

Fig. 3.6: Main factor Motivating usage of Biogas Plants

The major problem, however, lies with the inability of rural cattle farmers to afford the full cost of biogas plants which are very expensive. For example, in 2009 the average investment cost of a 10 m^3 digester ranged from GH¢ 4,000 to 6,000 ($ 2,800 and 4,200). These figures are far above the financial capability of the rural farmer and it is imperative that special microcredit schemes are developed as a means of promoting biogas plants among poor farmers.

FEED MATERIAL USED IN BIOGAS PLANTS

It has been shown already that the majority of households and institutions patronize biogas plants mainly as waste (nightsoil) treatment systems and as replacement for septic tanks. Most sanitary plants are either designed to treat nightsoil from flushing toilets or nightsoil from KVIP toilets. Apart from nightsoil, slaughterhouse wastes have also been treated using biogas plants. However, due to the high fibre content of slaughterhouse wastes, most plants constructed to handle the waste have broken down due to the formation of scum, poor operation and maintenance, and inappropriate siting of plants, as observed Tepa slaughterhouse. Scum is formed when fibrous feed materials accumulate at the top of the slurry, thicken and harden, thus obstructing the flow of gas into the gasholder. The gas is then forced to escape into the atmosphere via the inlet and outlet pipes of the digester. Figure gives an idea about the composition of feeding materials used in biogas plants.

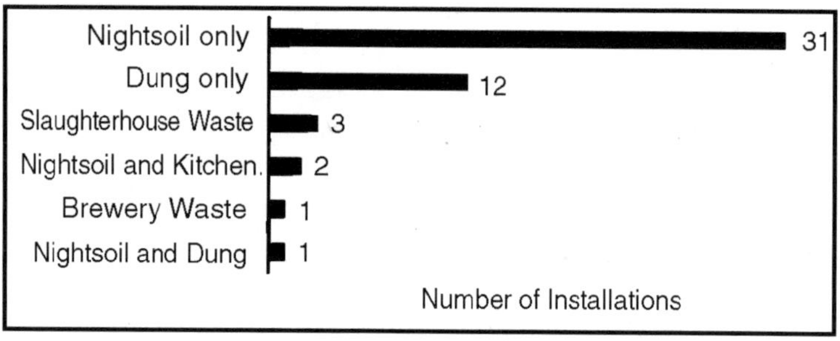

Fig. 3.7: Feed materials used in biogas plants in Ghana

TYPES AND SIZES OF PLANTS DISSEMINATED

With the exception of an Upflow Anaerobic Sludge Blanket (UASB) plant constructed at Guinness Ghana Breweries Limited in Kumasi, most of the plants surveyed are either floating-drum or fixed-dome digesters.

All the floating-drum digesters are of the water-jacket type with a spherical digester (BORDA model) while the gasholders are fabricated from mild steel apart from two glass-fibre gasholders at Tepa slaughterhouse and Holy Family hospital, Nkawkaw, and a high density polythene gasholder at GIMPA (Ghana Institute of Management and Professional Administration).

Despite the merits of the floating-drum digester, the fixed-dome digester has found favour among most biogas service providers, accounting for more than 80% of installations sampled. Figure highlights the main types of biodigesters found in the survey. Fixed-dome plants are less expensive compared to floating-drum digesters. Fixed-dome models disseminated by the various companies are the CAMARTEC

(Centre for Agricultural Mechanization and Rural Technology) model, the Deenbandhu model, the Chinese dome model, and lately the Puxin digester.

It should, however, be emphasized that most of the biogas companies have also disseminated designs with slight variations from the aforementioned models just to suit the topographic conditions of a particular area.

The sizes of surveyed plants (digester and gasholder as one unit) range from 10 to 100m^3.Most household and institutional plants disseminated have volumes of 10m^3 and 50 m^3 respectively.

Series (tandem) digesters have also been constructed. The most important criterion in sizing the digester is to be able to meet the minimum retention time; this is very important if the plant is to treat human excreta.

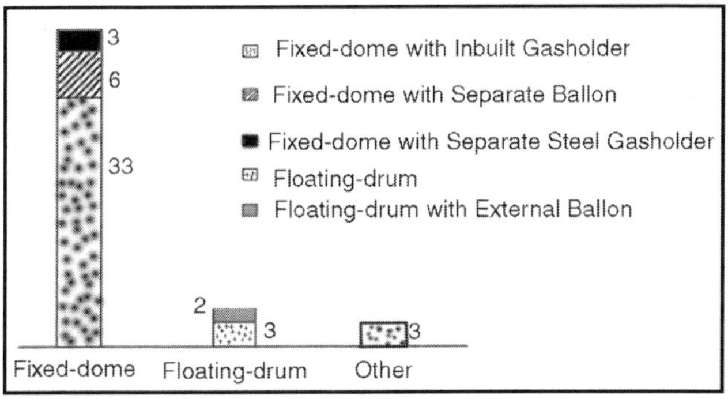

Fig. 3.8: Main types of Biodigesters Surveyed

Before the designer settles on the digester volume to use, he needs to consider factors such as daily availability of feed material and the retention time. The choice of the retention time in turn is influenced by the type of feed material, the digestion temperature, the plant type, and financial capabilities of the user, among others. Table gives the retention times used by some major biogas companies.

Table. Retention Time used by some Biogas Service Providers Under Mesophilic (25–30°C) Conditions

Biogas Service Provider	Retention Time [days]
Biogas Technology West Africa Limited (BTWAL)	30-60
Biosanitation Company Limited (BCL)	20-30
Beta Civil Engineering Limited (BCEL)	15-30
Institute of Industrial Research (IIR)	30-40
Biogas Engineering Limited (BEL)	30-60

If a biogas plant is designed to treat waste such as nightsoil, then the retention time should be long enough to allow for effective treatment of the waste, such that the effluent do not pose a health threat to the environment. The average retention time must be high enough to enable the degradation of the biomass; on the other hand, it must be kept as low as possible, because a high retention time means a high investment cost. Considering the high pathogen levels of human excreta, it is obvious that retention time of 30 days or less may not be adequate to kill enough pathogens under mesophilic digestion. According to Sasse *et al*, plants less than 30 m^3 should not be used for treating blackwater from flushing toilets because of the danger of reducing the retention time. It has also been recommended that a retention time between 70 and 80 days is needed for plants handling nightsoil, in order to ensure the reasonable destruction of pathogens.

As seen in Table, some service providers use a retention time lower than 30 days for designing plants to digest nightsoil, which is inappropriate and should be a matter of concern to environmentalists. However, BCL, BCEL, IIR and BTWAL have been installing systems that have post-treatment facilities ranging from filtration tanks to solar concentrators for further treatment of the effluent. There is still the need to conduct a thorough analysis of effluent samples to determine the safety of the 'treated' effluent.

TANDEM DIGESTERS

Tandem plants have two or more digesters working in series such that the effluent of one digester becomes the feed material for the second digester and so on. They may comprise two or more of the same type of digester or different digesters working together as one unit. The main advantages of tandem digesters (two-stage systems) over one digester (one-stage) systems include higher biomass conversion efficiency, higher methane concentration in the gas, better process reliability and stability, and high quality organic fertilizer. Despite the aforementioned merits, tandem systems are not widely disseminated worldwide because they are expensive and difficult to engineer,

construct and operate. Tandem plants have been constructed to treat sewage by BTWAL, BCEL, IIR, and BCL. Most common tandem digesters found consist of two series digesters except the plant at Pope John's Seminary in Koforidua and Hebron Prayer camp at Hebron, where BCEL has constructed systems comprising four 10 m^3 Puxin digesters. The main problem with tandem digester designs used by service providers pertains to the inappropriate way of connecting adjacent digesters via a horizontal pipe near the base of the digesters.

This arrangement leads to inefficient digestion since the second digester merely serves as an expansion chamber of the first digester rather than serving as a digester on its own. Moreover, digestion in the first digester will also be inefficient as slurry flows freely into the second digester. Figure shows the layout of tandem digesters as used by biogas service providers as opposed to the recommended layout.

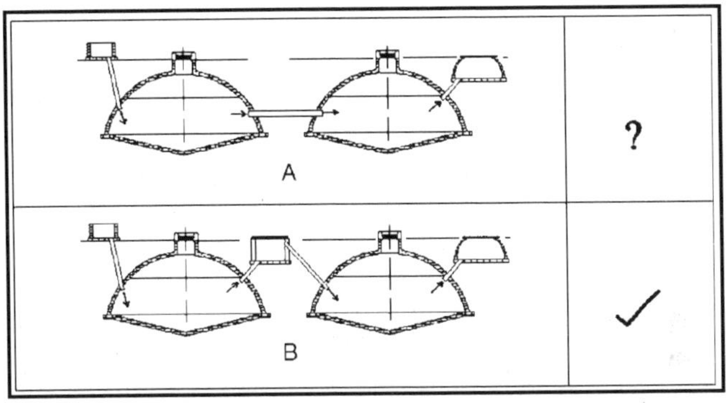

Fig. 3.9: (A) Tandem Digester Layout used by BTWAL, BCEL, and BCL, and (B) Appropriate and recommended Layout.

Tandem plants require a lot of calculations before designing them. The designs with horizontal pipes connecting adjacent digesters can work efficiently only if valves are placed at vital points to control the movement of fermentation slurry from one digester to another, and also the flow of effluent from the last digester. However, such an arrangement will be labour intensive in terms of operation and will also require skilled attendants.

The correct arrangement of tandem plants, is to connect a pipe near the base of the first digester to an expansion chamber which empties into the next digester. This arrangement ensures both plants function as fixed-dome systems, and ensures the effluent from the last digester undergoes thorough digestion.

COMMUNITY BIOSANITATION PROJECTS–BIOLATRINES (BIOTOILETS)

A total of 9 biolatrines were surveyed. Of the 9 biolatrines visited, four were not functioning due to design and operational challenges. The major problem with the design of biolatrines pertains to the connection between the toilet seat and the inlet pipe of the digester. If faeces got stucked to the inlet pipe, aerobic digestion takes place and bad odour is generated. This is believed to be the cause of breakdown of biolatrines at Kaase and Parkoso in the Ashanti Region.

Even though, experience from the Kaase and Parkoso project have led to improved designs, there are still problems with gas leakage through the inlet pipe into the toilet house, as observed in the biolatrines at Abeman, Asokore- Mampong, Kotoku, and Tamale West hospital. Another major challenge concerns the mode of operation of the biolatrine. Even though, biolatrines are designed to handle faeces with little or no water, they rather perform efficiently if water is frequently used to flush the faeces.

Thus, most biolatrines having water reservoirs enable users to flush the toilet with water after use. The use of water regularly avoids the occurrence of aerobic decomposition of faeces entrained in the inlet pipe of the digeste. From the survey, two biolatrines were found to have been disseminated very close to rubbish dumps. Caretakers attributed low patronage of these biotoilets to unbearable stench from the dump.

It was also observed that communities benefiting from biolatrine projects were not involved in the planning and implementation of the projects. The absence of interaction between community members and service providers on the functions of biolatrines has created ignorance among communities on the importance of biolatrines in improving sanitation and health.

GAS UTILIZATION

Due to low level of dissemination of plants handling cow dung, especially after the Appolonia Electrification project in 1992, the focus of biogas technology shifted from provision of energy (use of gas) to improvement in sanitation (treatment of waste).

This development has created a situation where most plants have been constructed without adequate arrangements for the usage of the gas thereby leading to gas leakage.

Gas leakage is a serious problem. In addition to the discomfort and health implications caused by inhalation of the gas, there is the potential of fire outbreak since mixtures of biogas and air in certain concentrations (6–12% biogas) could be explosive and may cause damage to human life and property. Moreover leakage of biogas, which contains about 60% methane, into the atmosphere offsets environmental gains made by putting up the plant.

Out of the 32 functioning plants studied, 12 of them experienced gas leakage problems. At both the Holy Family and the St. Dominic hospitals, the balloon gasholder was completely deteriorated and gas leaked freely into the atmosphere. Gas flaring is regularly done in a few locations, including the biogas plants financed at AngloGold Ashanti in Obuasi, where caretakers have been trained to flare the gas at regular periods in order to avoid high pressures within the digester.

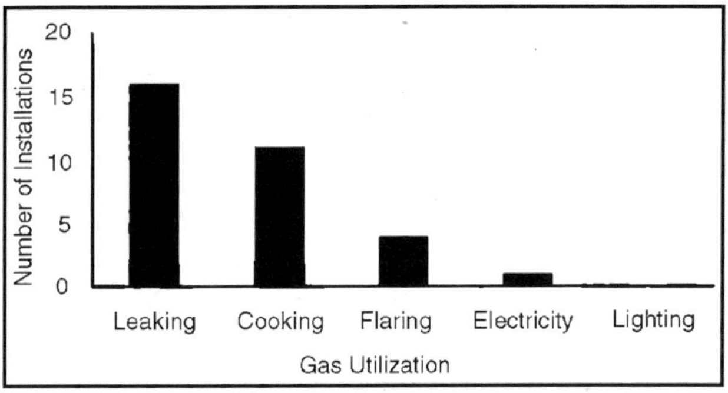

Fig. 3.10: Status of Biogas usage in Surveyed Plants (based on the number of Plants Functioning fully or Partially)

EFFLUENT (DIGESTED SLURRY) UTILIZATION

Biogas plants built with the ultimate aim of generating organic fertilizer are rare. The digested slurry is not seen as a resource and most plants discharge their effluent into the surrounding bush or the public drain. Some plants built by BCEL and BTWAL have systems designed to pump the treated effluent for reuse in flushing toilets.

The major concern pertains to the safety of the effluent discharged into the public drain. There is the need to conduct a comprehensive analysis of samples of the effluent in order to determine the efficiency of the digestion process of sanitary plants.

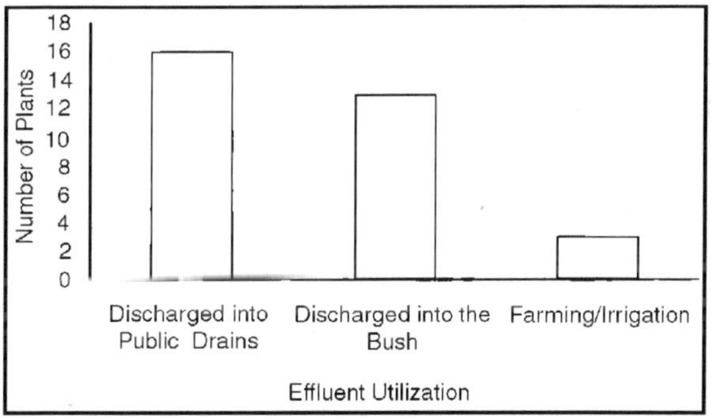

Fig. 3.11: Status of Effluent usage in Surveyed Plants (based on the number of Plants Functioning partially or fully)

The effluent from biolatrines is not used because of cultural attitude towards the use of human excreta as manure. In a study to assess the performance of biogas technology in Tanzania, Marree *et al* advocates for the use of the effluent from biolatrines as fertilizer in the cultivation of fodder grass. This approach, if adopted, will eliminate the stigma regarding the use of human manure in agriculture. Moreover, there is also the need to design and promote programmes that target farmers practicing mixed farming where the effluent can be easily and appropriately used as organic fertilizer.

OPERATION AND MAINTENANCE

A biogas plant may be well designed and constructed but still fail to perform efficiently if it is operated wrongly. Poor operation and maintenance could lead to the collapse of the plant. In a typical plant, good operational activities include, inter alia, daily feeding, cleaning of stoves and other appliances, checking of gas leakages especially at pipe joints and gas valves, and inspection of balloon gasholders. A well engineered plant will function properly as long as these tasks are performed reliably.

The lack of skilled attendants of biogas plants may have contributed more than any factor to the breakdown of most biogas plants. It was observed that users of biogas plants have little or no knowledge of the functions of the biogas plant. In addition, users were not sufficiently educated by the service companies. Of the 50 plants visited, only the installations (3 out of 24 plants were visited) at Anglogold Ashanti Limited were monitored daily by a trained engineer (caretaker); this has ensured continuous operation of the plants since 2001. Biogas plants disseminated in the Catholic hospitals visited have received some level of maintenance; caretakers of the plants had frequently repaired plant components such as the digester, the mixing tank, the expansion tank, and the steel gasholder in floating-drum plants. Unfortunately, balloon gasholders were found deteriorated at St.

Dominic and Holy Family hospitals at Akwatia and Nkawkaw, respectively; moreover, gas appliance were not been used since the gas escapes freely into the atmosphere through holes on the balloon. Other notable maintenance problems were found in biotoilet systems. Service providers do not spend enough time to take users through the rudiments and the functions of various components of the biogas plant.

Broken down plants were not repaired because owners were reluctant to spend additional financial resources on the upkeep of plants; this makes the floating-drum digester especially unfavourable due to the regular maintenance of the steel drum. Follow-up services were almost nonexistent. Service providers attributed this to the unavailability of funds to carry out follow-up programmes.

DEVELOPMENT OF A NATIONAL PROGRAMMES ON BIOGAS TECHNOLOGY

There is the need to develop a national biogas programmes with a three-pronged focus: agriculture (organic fertilizer production), sanitation, and energy. This national initiative must have distinct programmes for both domestic and institutional plants. Promotion of farm plants among livestock farmers or those interested in improving their yields will encourage the utilization of digested slurry in agriculture and reduce the dependence on imported inorganic fertilizers.

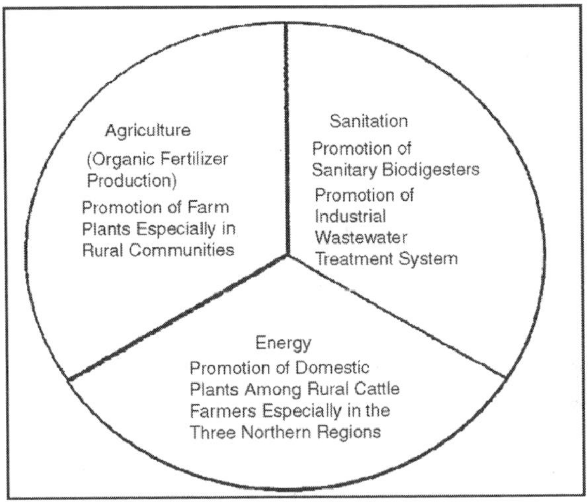

Fig. 3.12: Proposed Three-pronged Approach for Future Biogas Dissemination Programmes

The gas generated could be used for cooking, water heating, and lighting. Large-scale dissemination of sanitary plants will improve hygiene in both urban and rural areas.

Replacement of KVIPs with well-engineered biotoilets will ameliorate sanitation problems experienced in public toilets especially in communities where water is scarce. Moreover, the incidence of excretarelated diseases such as dysentery and cholera will be substantially minimized since micro- organisms that cause these diseases are eliminated during the anaerobic digestion process.

Large-scale local industries such as oil palm processing and brewing companies generate large volumes of wastewater that pollute the environmentwhen discharged untreated or partially treated. The use of anaerobic fermentation systems in the treatment of industrial wastewater will reduce pollution and improve the quality of the environment. For the programmes to be successful, government must provide funding to support biogas training and microfinance programmes targeted at social groups with high prospects for adopting biogas technology such as farmers in the three Northern Regions.

FORMATION OF A NATIONAL BIOGAS PROMOTION BODY

From the experience of biogas technology dissemination in developing countries such, Tanzania, China, India, and Nepal, it is obvious that biogas promotion cannot be left in the hands of private companies alone. The success story of the technology dissemination in China, India, and Nepal can be partly attributed to the direct involvement of the State through special national or parastatal bodies empowered to lead the campaign in biogas dissemination.

There is therefore the need to either set up a national body that is solely responsible for the promotion of biogas technology, or establish a dedicated unit within one of the existing energy organizations for the same purpose.

❐

4

Human Population and Environmental Issues

POPULATION GROWTH, VARIATION AMONG NATIONS

Our global human population, 6 billion at present, will cross the 7 billion mark by 2015. The needs of this huge number of human beings cannot be supported by the Earth's natural resources, without degrading the quality of human life. In the near future, fossil fuel from oil fields will run dry. It will be impossible to meet the demands for food from existing agro systems. Pastures will be overgrazed by domestic animals and industrial growth will create ever- greater problems due to pollution of soil, water and air.

Seas will not have enough fish. Larger ozone holes will develop due to the discharge of industrial chemicals into the atmosphere, which will affect human health. Global warming due to industrial gases will lead to a rise in sea levels and flood all low-lying areas, submerging coastal agriculture as well as towns and cities. Water 'famines' due to the depletion of fresh water, will create unrest and eventually make countries go to war. The control over regional biological diversity, which is vital for producing new medicinal and industrial products, will lead to grave economic conflicts between biotechnologically advanced nations and the biorich countries.

Degradation of ecosystems will lead to extinction of thousands of species, destabilizing natural ecosystems of great value. These are only some of the environmental problems related to an increasing human population and more intensive use of resources that we are likely to face in future. These effects can be averted by creating a mass environmental awareness movement that will bring about a change in people's way of life.

Increase in production per capita of agricultural produce at a global level ceased during the 1980's. In some countries, food shortage has become a permanent feature. Two of every three children in South Africa are underweight. In other regions famines due to drought have become more frequent. Present development strategies have not been able to successfully address these problems related to hunger and malnutrition.

On the other hand, only 15% of the world's population in the developed world is earning 79% of income! Thus the disparity in the extent of per capita resources that are used by people who live in a 'developed' country as against those who live in a 'developing' country is extremely large. Similarly, the disparity between the rich and the poor in India is also growing.

The increasing pressures on resources place great demands on the in- built buffering action of nature that has a certain ability to maintain a balance in our environment. However, current development strategies that essentially lead to short-term gains have led to a breakdown of our Earth's ability to replenish the resources on which we depend.

GLOBAL POPULATION GROWTH

The world population is growing by more than 90 million per year, of which 93% is in developing countries. This will essentially prevent their further economic 'development'. In the past, population growth was a gradual phenomenon and the Earth's ability to replenish resources was capable of adjusting to this increase. In the recent past, the escalation in growth of human numbers has become a major cause of our environmental problems. Present projections show that if our population growth is controlled, it will still grow to 7.27 billion by 2015.

However, if no action is taken it will become a staggering 7.92 billion. Human population growth increased from: 1 to 2 billion, in 123 years. 2 to 3 billion, in 33 years. 3 to 4 billion, in 14 years. 4 to 5 billion, in 13 years. 5 to 6 billion, in 11 years. It is not the census figures alone that need to be stressed, but an appreciation of the impact on natural resources of the rapid escalation in the rate of increase of human population in the recent past. The extent of this depletion is further increased by affluent societies that consume per capita more energy and resources, that less fortunate people.

This is of great relevance for developing a new ethic for a more equitable distribution of resources. In the first half of the 1900s human numbers were growing rapidly in most developing countries such as India and China. In some African countries the growth was also significant. In contrast, in the developed world population growth had slowed down. It was appreciated that the global growth rate was depleting the Earth's resources and was a direct impediment to human development.

Several environmental ill-effects were linked with the increasing population of the developing world. Poverty alleviation programs failed, as whatever was done was never enough as more and more people had to be supported on Earth's limited resources. In rural areas population growth led to increased fragmentation of farm land and unemployment. In the urban sector it led to inadequate housing and an increasing level of air pollution from traffic, water pollution from sewage, and an inability to handle solid waste. By the 1970s most countries in the developing world had realised that if they had to develop their economics and improve the lives of their citizens they would have to curtail population growth.

Though population growth shows a general global decline, there are variations in the rate of decline in different countries. By the 1990s the growth rate was decreasing in most countries such as China and India. The decline in the 90s was greatest in India. However, fertility continues to remain high in sub Saharan African countries. There are cultural, economic, political and demographic reasons that explain the differences in the rate of population control in different countries. It also varies in different parts of certain countries and is linked with community and/or religious thinking. Lack of Government initiatives for Family Welfare Programmes and a limited access to a full range of contraceptive measures are serious impediments to limiting population growth in several countries.

POPULATION EXPLOSION–FAMILY WELFARE PROGRAMMES

In response to our phenomenal population growth, India seriously took up an effective Family Planning Programmes which was renamed the Family Welfare Programmes. Slogans such as 'Hum do hamare do' indicated that each family should not have more than two children. It however has taken several decades to become effective. At the global level by the year 2000,600 million, or 57% of women in the reproductive age group, were using some method of contraception.

However the use of contraceptive measures is higher in developed countries–68%, and lower in developing countries-55%. Female sterilization is the most popular method of contraception used in developing countries at present. This is followed by the use of oral contraceptive pills and, intrauterine devices for women, and the use of condoms for men. India and China have been using permanent sterilization more effectively than many other countries in the developing world. The best decision for the method used by a couple depends on a choice that they make for themselves.

This must be based on good advice from doctors or trained social workers who can suggest the full range of methods available for them to choose from. Informing the public about the various contraceptive measures that are available is of primary importance. This must be done actively by Government Agencies such as Health and Family Welfare, as well as Education and Extension workers.

It is of great importance for policy makers and elected representatives of the people–Ministers, MPs, MLAs at Central and State levels–to understandnthe great and urgent need to support Family Welfare. The media must keep people informed about the need to limit family size and the ill effects of a growing population on the worlds resources. The decision to limit family size depends on a couple's background and education. This is related to Government Policy, the effectiveness of Family Welfare Programs, the educational level, and information levels in mass communication. Free access to Family Welfare information provided through the Health Care System, is in some cases unfortunately counteracted by cultural attitudes.

Frequently misinformation and inadequate information are reasons why a family does not go in for limiting its size. The greatest challenge the world now faces is how to supply its exploding human population with the resources it needs. It is evident that without controlling human numbers, the Earth's resources will be rapidly exhausted. In addition economically advanced countries and rich people in poorer countries use up more resources than they need.

As population expands further, water shortages will become acute. Soil will become unproductive. Rivers, lakes and coastal waters will be increasingly polluted. Water related diseases already kill 12 million people every year in the developing world. By 2025, there will be 48 countries that are starved for water. Air will become increasingly polluted.

Air pollution already kills 3 million people every year. The first 'green revolution' in the '60s produced a large amount of food but has led to several environmental problems. Now, a new green revolution is needed, to provide enough food for our growing population, that will not damage land, kill rivers by building large dams, or spread at the cost of critically important forests, grasslands and wetlands. The world's most populous regions are in coastal areas.

These are critical ecosystems and are being rapidly destroyed. Global climate change is now a threat that can affect the very survival of high population density coastal communities. In the sea, fish populations are suffering from excessive fishing.

Once considered an inexhaustible resource, over fishing has depleted stocks extremely rapidly. It will be impossible to support further growth in coastal populations on existing fish reserves. Human populations will inevitably expand from farm lands into the remaining adjacent forests. Many such encroachments in India have been regularised over the last few decades.

The extinction of plant and animal species resulting from shrinking habitats threatens to destroy the Earth's living web of life. Energy use is growing, both due to an increasing population, and a more energy hungry lifestyle that increasingly uses consumer goods that require large amounts of energy for their production, packaging, and transport. Our growing population also adds to the enormous amount of waste. With all these linkages between population growth and the environment, Family Welfare Programs have become critical to human existence.

Planning for the Future

How Governments and people from every community meet challenges such as limiting population size, protecting the natural environment, change their consumer oriented attitudes, reduce habits that create excessive waste, elevates poverty and creates an effective balance between conservation and development will determine the worlds future.

The Urban Challenge

Population increases will continue in urban centers in the near future. The UN has shown that by 2025 there will be 21 "megacities" most of which will be situated in developing countries. Urban centers are already unable to provide adequate housing, services such as water and drainage systems, growing energy needs, or better opportunities for income generation.

Methods of Sterilization

India's Family Welfare Programmes has been fairly successful but much still needs to be achieved to stabilize our population. The most effective measure is the one most suited to the couple once they have been offered all the various options that are available. The Family Welfare Programmes advocates a variety of measures to control population.

Permanent methods or sterilisation are done by a minor surgery. Tubectomy in females is done by tying the tubes that carry the ovum to the uterus. Male sterilization or vasectomy, is done by tying the tubes that carry the sperm. Both are very simple procedures, done under local anesthesia, are painless and patients have no post operative problems.

Vasectomy does not cause any loss in the male's sexual ability but only arrests the discharge of sperm. There are several methods of temporary birth control.

Condoms are used by males to prevent sperms from fertilizing the ovum during intercourse. Intrauterine devices (Copper Ts) are small objects which can be placed by a doctor in the uterus so that the ovum cannot be implanted, even if fertilized.

They do not disturb any functions in the woman's life or work. Oral contraceptive tablets (pills) and injectable drugs are available that prevent sperms from fertilizing the ovum. There are also traditional but less reliable methods of contraception such as abstinence of the sexual act during the fertile period of the women's cycle and withdrawal during the sexual act.

Urbanization

In 1975 only 27% of the people in the developing world lived in urban areas. By 2000 this had grown to 40% and by 2030 well informed estimates state that this will grow to 56%. The developed world is already highly urbanized with 75% of its population living in the urban sector.

Urban Environments

Nearly half the world's population now lives in urban areas. The high population density in these areas leads to serious environmental issues. Today, more than 290 million people live in towns and cities in India. There were 23 metros in India in 1991, which grew to 40 by 2001. Urban population growth is both due to migration of people to towns and cities from the rural sector in search of better job options as well as population growth within the city.

As a town grows into a city it not only spreads outwards into the surrounding agricultural land or natural areas such as forests, grasslands and wetlands but also grows skywards with high rise buildings. The town also loses its open spaces and green cover unless these are consciously preserved. This destroys the quality of life in the urban area. Good urban planning is essential for rational landuse planning, for upgrading slum areas, improving water supply and drainage systems, providing adequate sanitation, developing effective waste water treatment plants and an efficient public transport system.

Unplanned and haphazard growth of urban complexes has serious environmental impacts. Increasing solid waste, improper garbage disposal and air and water pollution are frequent side effects of urban expansions.

While all these issues appear to be under the preview of local Municipal Corporations, better living conditions can only become a reality if every citizen plays an active role in managing the environment. This includes a variety of "Dos and Don'ts" that should become an integral part of our personal lives.

Apart from undertaking actions that support the environment every urban individual has the ability to influence a city's management. He or she must see that the city's natural green spaces, parks and gardens are maintained, river and water fronts are managed appropriately, roadside tree cover is maintained, hill slopes are afforested and used as open spaces and architectural and heritage sites are protected. Failure to do this leads to increasing urban problems which eventually destroys a city's ability to maintain a healthy and happy lifestyle for its dwellers. All these aspects are closely linked to the population growth in the urban sector. In many cities growth outstrips the planner's ability to respond to this in time for a variety of reasons.

Mega cities in India	Population (in millions) in 2001	Projection (in millions) for 2015
Mumbai	16.5	22.6
Kolkata	13.3	16.7
Delhi	13.0	20.9

Small urban centers too will grow rapidly during the next decades and several rural areas will require reclassification as urban centers. India's urban areas will grow by a projected 297 million residents. In India people move to cities from rural areas in the hope of getting a better income. This is the 'Pull' factor. Poor opportunities in the rural sector thus stimulates migration to cities. Loss of agricultural land to urbanisation and industry, the inability of governments to sustainably develop the rural sector, and a lack of supporting infrastructure in rural areas, all push people from the agricultural and natural wilderness ecosystems into the urban sector. As our development strategies have focused attention mostly on rapid industrial development and relatively few development options are offered for the agricultural rural sector, a shift of population is inevitable.

As population in urban centers grows, they draw on resources from more and more distant areas. The "Ecological footprint" corresponds to the land area necessary to supply natural resources and disposal of waste of a community. At present the average ecological footprint of an individual at the global level is said to be 2.3 hectares of land per capita.

But it is estimated that the world has only 1.7 hectares of land per individual to manage these needs sustainably. This is thus an unsustainable use of land. The pull factor of the urban centers is not only due to better job opportunities, but also better education, health care and relatively higher living standards. During the last few decades in India, improvements in the supply of clean water, sanitation, waste management, education and health care has all been urban centric, even though the stated policy has been to support rural development. Thus in reality, development has lagged behind in the rural sector that is rapidly expanding in numbers. For people living in wilderness areas in our forests and mountain regions, development has been most neglected. It is not appropriate to use the development methods used for other rural communities for tribal people who are dependent on collecting natural resources from the forests. A different pattern of development that is based on the sustainable extraction of resources from their own surroundings would satisfy their development aspirations. In general the growing human population in the rural sector will only opt to live where they are if they are given an equally satisfying lifestyle.

The Wilderness–Rural-urban Linkage

The environmental stresses caused by urban individuals covers an 'ecological footprint' that goes far beyond what one expects. The urban sector affects the land at the fringes of the urban area and the areas from which the urban center pulls in agricultural and natural resources. Urban centers occupy 2% of the worlds' land but use 75% of the industrial wood. About 60% of the world's water is used by urban areas of which half irrigates food crops for urban dwellers, and one third goes to industry and the rest is used for household use and drinking water.

The impact that urban dwellers have on the environment is not obvious to them as it happens at distant places which supports the urban ecosystem with resources from agricultural and even more remote wilderness ecosystems.

Urban Poverty and the Environment

The number of poor people living in urban areas is rapidly increasing. A third of the poor people in the world live in urban centers. These people live in hutments in urban slums and suffer from water shortages and unsanitary conditions. In most cases while a slum invariably has unhygenic surroundings, the dwellings themselves are kept relatively clean. It is the 'common' areas used by the community that lacks the infrastructure to maintain a hygienic environment. During the 1990s countries that have experienced an economic crisis have found that poor urban dwellers have lost their jobs due to decreasing demands for goods, while food prices have risen. Well paid and consistent jobs are not as easily available in the urban centers at present as in the past few decades. One billion urban people in the world live in inadequate housing, mostly in slum areas, the majority of which are temporary structures.

However, low income groups that live in high rise buildings can also have high densities and live in poor unhygienic conditions in certain areas of cities. Illegal slums often develop on Government land, along railway tracks, on hill slopes, riverbanks, marshes, etc. that are unsuitable for formal urban development. On the riverbanks floods can render these poor people homeless. Adequate legal housing for the urban poor remains a serious environmental concern.

Urban poverty is even more serious than rural poverty, as unlike the rural sector, the urban poor have no direct access to natural resources such as relatively clean river water, fuelwood and non wood forest products. The urban poor can only depend on cash to buy the goods they need, while in the rural sector they can grow a substantial part of their own food. Living conditions for the urban poor are frequently worse than for rural poor. Both outdoor and indoor air pollution due to high levels of particulate matter and sulphur dioxide from industrial and vehicle emissions lead to high death rates from respiratory diseases. Most efforts are targeted at outdoor air pollution. Indoor air pollution due to the use of fuel wood, waste material, coal, etc. in 'chulas' is a major health issue. This can be reduced by using better designed 'smokeless' chulas, hoods and chimneys to remove indoor smoke. With the growing urban population, a new crisis of unimaginable proportions will develop in the next few years. Crime rates, terrorism, unemployment, and serious environmental health related issues can be expected to escalate. This can only be altered by stabilizing population growth on a war footing.

ENVIRONMENT AND HUMAN HEALTH

Environment related issues that affect our health have been one of the most important triggers that have led to creating an increasing awareness of the need for better environmental management. Changes in our environment induced by human activities in nearly every sphere of life have had an influence on the pattern of our health. The assumption that human progress is through economic growth is not necessarily true.

We expect urbanization and industrialization to bring in prosperity, but on the down side, it leads to diseases related to overcrowding and an inadequate quality of drinking water, resulting in an increase in waterborne diseases such as infective diarrhoea and air borne bacterial diseases such as tuberculosis. High-density city traffic leads to an increase in respiratory diseases like asthma. Agricultural pesticides that enhanced food supplies during the green revolution have affected both the farm worker and all of us who consume the produce.

Modern medicine promised to solve many health problems, especially associated with infectious diseases through antibiotics, but bacteria found ways to develop resistant strains, frequently even changing their behaviour in the process, making it necessary to keep on creating newer antibiotics. Many drugs have been found to have serious side effects. At times the cure is as damaging as the disease process itself. Thus development has created several long-term health problems. While better health care has led to longer life spans, coupled with a lowered infant mortality, it has also led to an unprecedented growth in our population which has negative implications on environmental quality. A better health status of society will bring about a better way of life only if it is coupled with stabilising population.

Environmental Health

Environmental health, as defined by WHO, comprises those aspects of human health, including quality of life, that are determined by physical, chemical, biological, social, and psychosocial factors in the environment. It also refers to the theory and practice of assessing, correcting, controlling, and preventing those factors in the environment that adversely affect the health of present and future generations. Our environment affects health in a variety of ways.

Climate and weather affect human health. Public health depends on sufficient amounts of good quality food, safe drinking water, and adequate shelter. Natural disasters such as storms, hurricanes, and floods still kill many people every year. Unprecedented rainfall trigger epidemics of malaria and water borne diseases. Global climate change has serious health implications. Many countries will have to adapt to uncertain climatic conditions due to global warming. As our climate is changing, we may no longer know what to expect.

There are increasing storms in some countries, drought in others, and a temperature rise throughout the world. The El Niño winds affect weather worldwide. The El Niño event of 1997/98 had serious impacts on health and well-being of millions of people in many countries. It created serious drought, floods, and triggered epidemics. New strategies must be evolved to reduce vulnerability to climate variability and changes.

Economic inequality and environmental changes are closely connected to each other. Poor countries are unable to meet required emission standards to slow down climate change. The depletion of ozone in the stratosphere (middle atmosphere) also has an important impact on global climate and in turn human health, increasing the amount of harmful ultraviolet radiation that reaches the Earth's surface. This results in diseases such as skin cancer.

Bhopal Gas Tragedy

The siting of industry and relatively poor regulatory controls leads to ill health in the urban centers. Accidents such as the Bhopal gas tragedy in 1984 where Union Carbide's plant accidentally released 30 tones of methyl isocyanate, used in the manufacture of pesticides, led to 3,330 deaths and 1.5 lakh injuries to people living in the area. Development strategies that do not incorporate ecological safeguards often lead to ill health. Industrial development without pollution control and traffic congestion affect the level of air pollution in many cities. On the other hand, development strategies that can promote health invariably also protect the environment. Thus environmental health and human health are closely interlinked. An improvement in health is central to sound environmental management. However this is rarely given sufficient importance in planning development strategies.

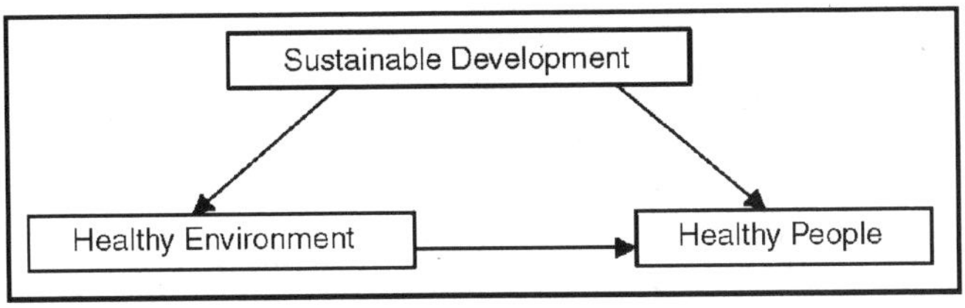

Examples of the Linkages

- Millions of children die every year due to diarrhoea from contaminated water or food. An estimated 2000 million people are affected by these diseases and more than 3 million children die each year from waterborne diseases across the world. In India, it is estimated that every fifth child under the age of 5 dies due to diarrhoea. This is a result of inadequate environmental management and is mainly due to inadequate purification of drinking water. Wastewater and/or sewage entering water sources without being treated leads to continuous gastrointestinal diseases in the community and even sporadic large epidemics. Large numbers of people in tropical countries die of malaria every year and millions are infected. An inadequate environmental management of stagnant water, which forms breeding sites of Anopheles mosquitoes is the most important factor in the spread of malaria. The resurgence of malaria in India is leading to cerebral malaria that affects the brain and has a high mortality.
- Millions of people, mainly children, have poor health due to parasitic infections, such as amoebiasis and worms. This occurs from eating infected food, or using poor quality water for cooking food. It is estimated that 36% of children in low-income countries and 12% in middle income countries are malnourished. In India, about half the children under the age of four are malnourished and 30% of newborns are significantly underweight.
- Hundreds of millions of people suffer serious respiratory diseases, including lung cancer and tuberculosis, from crowded homes and public places. Motor vehicle exhaust fumes, industrial fumes, tobacco smoke and cooking food on improper 'chulas', contribute to respiratory diseases.
- Millions of people are exposed to hazardous chemicals in their workplace or homes that lead to ill health due to industrial products where controls are not adhered to.
- Tens of thousands of people in the world die due to traffic accidents due to inadequate management of traffic conditions. Poor management at the accident site, and inability to reach a hospital within an hour causes a large number of deaths, especially from head injuries.

- Basic environmental needs such as clean water, clean air and adequate nutrition which are all related to environmental goods and services do not reach over 1000 million people living in abject poverty.
- Several million people live in inadequate shelters or have no roof over their heads especially in urban settings. This is related to high inequalities in the distribution of wealth and living space.
- Population growth and the way resources are being exploited and wasted, threatens environmental integrity and directly affects health of nearly every individual.
- Health is an outcome of the interactions between people and their environment. Better health can only come from a more sustainable management of the environment.

Important Strategic Concerns

- The world must address people's health care needs and sustainable use of natural resources, which are closely linked to each other.
- Strategies to provide clean pottable water and nutrition to all people is an important part of a healthy living environment.
- Providing clean energy sources that do not affect health is a key to reducing respiratory diseases.
- Reducing environmental consequences of industrial and other pollutants such as transport emissions can improve the status of health.
- Changing patterns of agriculture away from harmful pesticides, herbicides and insecticides which are injurious to the health of farmers and consumers by using alternatives such as Integrated Pest Management and non-toxic biopesticides can improve health of agricultural communities, as well as food consumers.
- Changing industrial systems into those that do not use or release toxic chemicals that affect the health of workers and people living in the vicinity of industries can improve health and environment.
- There is a need to change from using conventional energy from thermal power that pollutes air and nuclear power that can cause serious nuclear disasters to cleaner and safer sources such as solar, wind and ocean power that do not affect human health. Providing clean energy is an important factor that can lead to better health.
- The key factors are to control human population and consume less environmental goods and services which could lead to 'health for all'. Unsustainable use of resources by an ever growing

population leads to unhealthy lives. Activities that go on wasting environmental goods and destroying its services by producing large quantities of non degradable wastes, leads to health hazards.

- Poverty is closely related to health and is itself a consequence of improper environmental management. An inequitable sharing of natural resources and environmental goods and services, is linked to poor health. The world's consumption of non-renewable resources is concentrated in the developed countries. Rich countries consume 50 times more per capita than people in less developed countries. This means that developed countries also generate proportionately high quantities of waste material, which has serious health concerns.

Definition of Health Impact Assessment (HIA) by WHO

Health impact assessment is a combination of procedures, methods and tools by which a policy, programmes or project may be judged as to its potential effects on the health of a population, and the distribution of those effects within the population.

Climate and Health

Human civilizations have adapted mankind to live in a wide variety of climates. From the hot tropics to the cold arctic, in deserts, marshlands and in the high mountains. Both climate and weather have a powerful impact on human life and health issues. Natural disasters created by extremes of weather (heavy rains, floods, hurricanes) which occur over a short period of time, can severely affect health of a community. Poor people are more vulnerable to the health impacts of climate variability than the rich.

Of approximately 80,000 deaths which occur world-wide each year as a result of natural disasters about 95% are in poor countries. In weather-triggered disasters hundreds of people and animals die, homes are destroyed, crops and other resources are lost. Public health infrastructure, such as sewage disposal systems, waste management, hospitals and roads are damaged. The cyclone in Orissa in 1999 caused 10,000 deaths. The total number of people affected was estimated at 10 to 15 million!

Human physiology can adapt to changes in weather, within certain limits. However, marked short-term fluctuations in weather lead to serious health issues. Heat waves cause heat-related illness and death (*e.g.*, heat stroke). The elderly and persons with existing heart or respiratory diseases are more vulnerable. Heat waves in India in 1998 were associated with many deaths.

Climate plays an important role in vector-borne diseases transmitted by insects such as mosquitoes. These disease transmitters are sensitive to direct effects of climate such as temperature, rainfall patterns and wind.

Climate affects their distribution and abundance through its effects on host plants and animals. Malaria transmission is particularly sensitive to weather and climate. Unusual weather conditions, for example a heavy downpour, can greatly increase the mosquito population and trigger an epidemic.

In the desert and at highland fringes of malarious areas, malaria transmission is unstable and the human population lacks inherent protective immunity. Thus, when weather conditions (rainfall and temperature) favour transmission, serious epidemics occur in such areas. Fluctuations in malaria over the years have been linked to changes in rainfall associated with the El Niño cycle.

Infectious Diseases

Many infectious diseases have re-emerged with a vengeance. Loss of effective control over diseases such as malaria and tuberculosis, have led to a return of these diseases decades after being kept under stringent control. Other diseases were not known to science earlier and seem to have suddenly hit our health and our lives during the last few decades. AIDS, due to the Human Immunodeficiency Virus (HIV) caused through sexual transmission and Severe Acute Respiratory Syndrome (SARS) are two such examples.

While these cannot be directly related to environmental change, they affect the environment in which we live by forcing a change in lifestyles and behaviour patterns. For example the SARS outbreak prevented people from several countries from traveling to other countries for months, severely affecting national economies, airline companies and the tourism industry. Why have infectious diseases that were related to our environment that were under control suddenly made a comeback? Diseases such as tuberculosis have been effectively treated with anti-tubercular drugs for decades.

These antibiotics are used to kill off the bacteria that causes the disease. However nature's evolutionary processes are capable of permitting bacteria to mutate by creating new genetically modified strains. Those that change in a way so that they are not affected by the routinely used antibiotics begin to spread rapidly. This leads to a reemergence of the disease. In the case of tuberculosis this has led to multi-drug resistant tuberculosis. This is frequently related to HIV which reduces an individual's immunity to bacteria such as mycobacterium tuberculosis that causes tuberculosis. The newer broad- spectrum antibiotics, antiseptics, disinfectants, and vaccines once thought of as the complete answer to infectious diseases have thus failed to eradicate infectious diseases. Experts in fact now feel that these diseases will be the greatest killers in future and not diseases such as malignancy or heart disease.

While antibiotic resistance is a well-known phenomenon there are other reasons for the reemergence of diseases. Overcrowding due to the formation of slums in the urban setting leads to several health hazards, including easier spread of respiratory diseases. Inadequate drinking water quality and poor disposal of human waste due to absence of a closed sewage system and poor garbage management are all urban health issues. This has led to a comeback of diseases such as cholera and an increased incidence of diarrhea and dysentery as well as infectious hepatitis (jaundice). With increasing global warming disease patterns will continue to change. Tropical diseases spread by vectors such as the mosquito will undoubtedly spread malaria further away from the equator. Global warming will also change the distribution of dengue, yellow fever, encephalitis, etc.

Warmer wetter climates could cause serious epidemics of diseases such as cholera. El Nino which causes periodic warming is likely to affect rodent populations. This could bring back diseases such as the plague. Globalisation and infectious disease Globalization is a world-wide process which includes the internationalization of communication, trade and economic organization. It involves parallel changes such as rapid social, economic and political adjustments. Whilst globalization has the potential to enhance the lives and living standards of certain population groups, for poor and marginalized populations in both the non-formal as well as formal economic sectors of developing countries, globalization enhances economic inequalities.

Tuberculosis

Tuberculosis (TB) kills approximately 2 million people each year. In India the disease has reemerged and is now more difficult to treat. A global epidemic is spreading and becoming more lethal. The spread of HIV/AIDS and the emergence of multidrug-resistant tuberculosis is contributing to the increasing morbidity of this disease. In 1993, the World Health Organization (WHO) declared that tuberculosis had become a global emergency. It is estimated that between 2002 and 2020, approximately 1000 million people will be newly infected, over 150 million people will get sick, and 36 million will die of TB–if its control is not rapidly strengthened. TB is a contagious disease that is spread through air.

Only people who are sick with pulmonary TB are infectious. When infectious people cough, sneeze, talk or spit, they emit the tubercle bacilli into the air. When a healthy person inhales these, he gets infected by the disease. Symptoms include prolonged fever, coughing spells and weight loss. It is estimated that, left untreated, each patient of active tuberculosis will infect on an average between 10 to 15 people every year. But people infected with TB will not necessarily get sick with the disease.

The immune system can cause the TB bacilli, which is protected by a thick waxy coat, to remain dormant for years. When an individual's immune system is weakened, the chances of getting active TB are greater.

- Nearly 1% of the world's population is newly infected with TB each year.
- It is estimated that overall, one third of the world's population is likely to be infected with the tuberculosis bacillus at some point in time.
- Five to ten per cent of people who are infected with TB (but who are not infected with HIV) become sick or infectious at some time during their life. (WHO, 2002). Factors Contributing to the rise in tuberculosis
- TB kills about 2 million people each year (including persons infected with HIV).
- More than 8 million people become sick with TB each year, one person in the world every second!
- About 2 million TB cases per year occur in sub-Saharan Africa. This number is rising rapidly as a result of the HIV/AIDS epidemic.
- Around 3 million TB cases per year occur in South-east Asia.
- Over a quarter of a million TB cases per year occur in Eastern .Europe

Tuberculosis in India

There are 14 million TB patients in India, account for one third of the global cases of TB. Everyday 20,000 Indians contract TB and more than 1,000 die due to this chronic illness. TB attacks working adults in the age group of 15 to 50 years.

HIV is accelerating the spread of TB

The link between HIV and TB affects a large number of people, each disease speeding the other's progress. HIV weakens the immune system. Someone who is HIV-positive and infected with TB is many times more likely to become seriously sick with TB rather than someone infected with TB who is HIV-negative. Tuberculosis is a leading cause of death among people who are HIV-positive, accounting for about 11% of AIDS deaths worldwide.

Poorly Managed TB programs are Threatening to make TB Incurable

Until 50 years ago, there were no drugs to cure tuberculosis. Now, strains that are resistant to one or more anti-TB drugs have emerged. Drugresistant tuberculosis is caused by inconsistent or partial treatment, when patients do not take all their drugs regularly for the required period, when doctors or health workers prescribe inadequate treatment regimens or where the drug supply is unreliable. From a public health perspective, poorly supervised or incomplete treatment of TB is worse than no treatment at all. When people fail to complete standard treatment regimens, or are given the wrong treatment regimen, they may remain infectious. The bacilli in their lungs may develop resistance to anti-TB drugs. People they infect will have the same drug-resistant strain. While drug-resistant TB is treatable, it requires extensive chemotherapy that is often very expensive and is also more toxic to patients.

Malaria

Malaria is a life-threatening parasitic disease transmitted by mosquitoes. The cause of malaria, a single celled parasite called plasmodium, was discovered in 1880.Later it was found that the parasite is transmitted from person to person through the bite of a female Anopheles mosquito, which requires blood for the growth of her eggs. Today approximately 40% of the world's population, mostly those living in the world's poorest countries, risk getting malaria.

The disease was once more widespread but it was successfully eliminated from many countries with temperate climates during the mid 20th century. Today malaria has returned and is found throughout the tropical and sub-tropical regions of the world and causes more than 300 million acute illnesses and at least one million deaths annually (WHO). There are several types of human malaria. Falciparum malaria is the most dangerous type of infection and is most common in Africa south of the Sahara, where it accounts for extremely high mortality rates. The malaria parasite enters the human host when an infected Anopheles mosquito bites an individual. Inside the human host, the parasite undergoes a series of changes as part of its complex life-cycle.

Its various stages allow plasmodia to evade the immune system, infect the liver and red blood cells, and finally develop into a form that is able to infect a mosquito again when it bites an infected person. Inside the mosquito, the parasite matures until it reaches the sexual stage where it can again infect a human host when the mosquito takes her next blood meal, 10 or more days later. Malaria symptoms appear about 9 to 14 days after the mosquito bite, although this varies with different plasmodium species. Malaria produces high fever, headache, vomiting and body ache. If drugs are not available for treatment, or the parasites are resistant to them, the infection can progress rapidly to become life-threatening. Malaria can kill by infecting and destroying red blood cells (anaemia) and by clogging the capillaries that carry blood to the brain (cerebral malaria) or other vital organs.

Malaria parasites are developing unacceptable levels of resistance to drugs. Besides this, many insecticides are no longer useful against mosquitoes transmitting the disease. Good environmental management by clearing pools of stagnant water during the monsoons is effective in reducing the number of mosquitoes. Mosquito nets treated with insecticide reduce malaria transmission and child deaths.

Prevention of malaria in pregnant women, through measures such as Intermittent Preventive Treatment and the use of insecticide-treated nets (ITNs), results in improvement in maternal health, as well as infant health and survival. Prompt access to treatment with effective up-to-date medicines, such as artemisinin-based combination therapies (ACTs), saves lives. If countries can apply these and other measures on a wide scale and monitor them carefully, the burden of malaria on society will be significantly reduced.

Water-related Diseases

Water Supply, Sanitation and Hygiene Development

Among the main problems are a lack of priority given to this sector, lack of financial resources, erratic water supply and sanitation services, poor hygiene related behaviour patterns, and inadequate sanitation in public places such as schools, hotels, hospitals, health centers, etc. One of the most important aspects is a lack of environmental education and awareness that these disease processes are related to poor environment management in various sectors.

Providing access to sufficient quantities of safe water, the provision of facilities for a sanitary disposal of excreta, and introducing sound hygiene related behaviour can reduce the morbidity and mortality caused by these risk factors. Environmental Sanitation and Hygiene Development About 2.4 billion people globally live under highly unsanitary conditions. Poor hygiene and behaviour pattern increase the exposure to risk of incidence and spread of infectious diseases. Water improperly stored in homes is frequently contaminated by inadequate management at the household level. This can be easily reduced through education and awareness of how waterborne diseases are transmitted.

Health and Water Resources Development

An important aspect related to water-related diseases (in particular: water- related vectorborne diseases) is attributable to the way water resources are developed and managed. In many parts of the world the adverse health impacts of dam construction, irrigation development and flood control is related to increased incidence of malaria, Japanese encephalitis, schistosomiasis, lymphatic filariasis and other conditions.

Other health issues indirectly associated with water resources development include nutritional status, exposure to agricultural pesticides and their residues.

Water Borne Diseases

Arid areas with rapidly expanding populations are already facing a crisis over water. Conservation of water and better management is an urgent need. The demand and supply balance is a vital part of developing sustainable use of water. This is being termed the 'Blue Revolution' and needs Governments, NGOs and people to work together towards a better water policy at International, National, State, regional and local levels. Locally good watershed management is a key to solving local rural problems. Present patterns of development are water hungry and water wasters. They do not address pollution and overuse. The linkages between managing water resources and health issues are have not been prioritised as a major source of environmental problems that require policy change, administrative capacity building and an increased financial support.

There are 4 Major types of Water Related Diseases

1 *Water borne diseases*: These are caused by dirty water contaminated by human and animal wastes, especially from urban sewage, or by chemical wastes from industry and agriculture. Some of these diseases, such as cholera and typhoid, cause serious epidemics. Diarrhoea, dysentery, polio, meningitis, and hepatitis A and E, are caused due to improper drinking water. Improving sanitation and providing treated drinking water reduces the incidence of these diseases.

2 *Water based diseases*: Aquatic organisms that live a part of their life cycle in water and another part as a parasite in man, lead to several diseases. In India, guinea worm affects the feet. Round worms live in the small intestine, especially of children.

3 *Water related vector diseases*: Insects such as mosquitoes that breed in stagnant water spread diseases such as malaria and filariasis. Malaria that was effectively controlled in India, has now come back as the mosquitoes have become resistant to insecticides. In addition, anti-malarial drugs are now unable to kill the parasites as they have become resistant to drugs. Change in climate is leading to the formation of new breeding sites. Other vector born diseases in India include dengue fever and filariasis. Dengue fever carries a high mortality. Filariasis leads to fever and chronic swelling over the legs.

4 *Water scarcity diseases:* In areas where water and sanitation is poor, there is a high incidence of diseases such as tuberculosis, leprosy, tetanus, etc. which occur when hands are not adequately washed.

Arsenic in Drinking Water

Arsenic in drinking-water is a serious hazard to human health. It has attracted much attention since its recognition in the 1990s of its wide occurrence in wellwater in Bangladesh. It occurs less frequently in most other countries. The main source of arsenic in drinking water is arsenic-rich rocks through which the water has filtered. It may also occur because of mining or industrial activity in some areas. WHO has worked with other UN organizations to produce a state-of-the-art review on arsenic in drinking water.

Drinking water that is rich in arsenic leads to arsenic poisoning or arsenicosis. Excessive concentrations are known to occur in some areas. The health effects are generally delayed and the most effective preventive measure is supplying drinking water which is free of arsenic. Arsenic contamination of water is also due to industrial processes such as those involved in mining, metal refining, and timber treatment. Malnutrition may aggravate the effects of arsenic on blood vessels. Water with high concentrations of arsenic if used over 5 to 20 years, results in problems such as colour changes on the skin, hard patches on the palms and soles, skin cancer, cancers of the bladder, kidney and lung, and diseases of the blood vessels of the legs and feet. It may also lead to diabetes, high blood pressure and reproductive disorders. Natural arsenic contamination occurs in Argentina, Bangladesh, Chile, China, India, Mexico, Thailand and the United States. In China (in the Province of Taiwan) exposure to arsenic leads to gangrene, known as 'black foot disease'.

Long term solutions for prevention of arsenicosis is based on providing safe drinking-water:

- Deeper wells are often less likely to be contaminated.
- Testing of water for levels of arsenic and informing users.
- Monitoring by health workers-people need to be checked for early signs of arsenicosis-usually by observing skin problems in areas where arsenic in known to occur.
- Health education regarding harmful effects of arsenicosis and how to avoid them.

Arsenic Poisoning–Bangladesh

More than half the population of Bangladesh is threatened by high levels of arsenic found in drinking water. This could eventually lead to an epidemic of cancers and other fatal diseases. Rezaul Morol, a young Bangladeshi man, nearly died from arsenic poisoning caused by drinking arsenic-laden well- water for several years. The doctor advised Rezaul to stop drinking contaminated water and eat more protein-rich food such as fish. Since then Rezaul feels a lot better and is happy that his skin is healing.

Diarrhoea

Though several types of diarrhoea which give rise to loose motions and dehydration occur all over the world, this is especially frequently observed in developing countries. It causes 4% of all deaths. In another 5% it leads to loss of health. It is caused by gastrointestinal infections which kill around 2.2 million people globally each year. Most of these are children in developing countries. The use of contaminated water is an important cause of this group of conditions. Cholera and dysentery cause severe, sometimes life threatening and epidemic forms of these diseases.

Effects on Health

Diarrhoea is the frequent passage of loose or liquid stools. It is a symptom of various gastrointestinal infections. Depending on the type of infection, the diarrhoea may be watery (for example in cholera caused by vibrio cholera) or passed with blood and mucous (in dysentery caused by an amoeba, E Histolitica). Depending on the type of infection, it may last a few days, or several weeks. Severe diarrhoea can become life threatening due to loss of excessive fluid and electrolytes such as Sodium and Potassium in watery diarrhoea.

This is particularly fatal in infants and young children. It is also dangerous in malnourished individuals and people with poor immunity. The impact of repeated diarrhoea on nutritional status is linked in a vicious cycle in children. Chemical or non-infectious intestinal conditions can also result in diarrhoea.

Causes of Diarrhoea

Diarrhoea is caused by several bacterial, viral and parasitic organisms. They are mostly spread by contaminated water. It is more common when there is a shortage of clean water for drinking, cooking and cleaning. Basic hygiene is important in its prevention. Water contaminated with human feces surrounding a rural water source, or from municipal sewage, septic tanks and latrines in urban centers, are important factors in the spread of these diseases. Feces of domestic animals also contain microorganisms that can cause diarrhoea through water. Diarrhoea is spread from one individual to another due to poor personal hygiene. Food is a major cause of diarrhoea when it is prepared or stored in unhygienic conditions. Water can contaminate food such as vegetables during irrigation. Fish and seafood from polluted water is a cause of severe diarrhoea. The infectious agents that cause diarrhoea are present in our environment.

In developed countries where good sanitation is available, most people get enough safe drinking water. Good personal and domestic hygiene prevents this disease which is predominantly seen in the developing world. About 1 billion people do not have access to clean water sources and 2.4 billion have no basic sanitation. In Southeast Asia, diarrhoea is responsible for 8.5% of all deaths. In 1998, diarrhoea was estimated to have killed 2.2 million people, most of whom were under 5 years of age.

Interventions: Key measures to reduce the number of cases of diarrhoea include:

- Access to safe drinking water.
- Improved sanitation.
- Good personal and food hygiene.
- Health education about how these infections spread.

Key measures to treat diarrhoea include:

- Giving more fluids than usual, (oral rehydration) with salt and sugar, to prevent dehydration.
- Continue feeding.
- Consulting a health worker if there are signs of dehydration or other problems.

In rural India, during the last decade public education through posters and other types of communication strategies has decreased infant mortality due to diarrhoea in several States. Posters depicting a child with diarrhoea being given water, salt and sugar solution to reduce death from dehydration has gone a long way in reducing both a serious condition requiring hospitalisation and intravenous fluids, as well as mortality.

Risks Due to Chemicals in Food

Food contaminated by chemicals is a major worldwide public health concern. Contamination may occur through environmental pollution of the air, water and soil. Toxic metals, PCBs and dioxins, or the intentional use of various chemicals, such as pesticides, animal drugs and other agrochemicals have serious consequences on human health. Food additives and contaminants used during food manufacture and processing adversely affects health.

Diseases spread by food: Some foodborne diseases though well recognized, have recently become more common. For example, outbreaks of salmonellosis which have been reported for decades, has increased within the last 25 years. In the Western hemisphere and in Europe, Salmonella serotype Enteritidis (SE) has become a predominant strain. Investigations of SE outbreaks indicate that its emergence is largely related to consumption of poultry or eggs.

While cholera has devastated much of Asia and Africa for years, its reintroduction for the first time in almost a century on the South American continent in 1991 is an example of a well recognised infectious disease re- emerging in a region after decades. While cholera is often waterborne, many foods also transmit infection. In Latin America, ice and raw or underprocessed seafood are important causes for cholera transmission.

Infection with a specific type of Escherichia coli (E. coli) was first described in 1982. Subsequently, it has emerged rapidly as a major cause of bloody diarrhoea and acute renal failure. The infection is sometimes fatal, particularly in children. Outbreaks of infection, generally associated with beef, have been reported in Australia, Canada, Japan, United States, in various European countries, and in southern Africa. Outbreaks have also implicated alfalfa sprouts, unpasteurized fruit juice, lettuce, game meat (meat of wild animals) and cheese curd.

In 1996, an outbreak of Escherichia coli in Japan affected over 6,300 school children and resulted in 2 deaths. Listeria monocytogenes (Lm): The role of food in the transmission of this condition has been recognized recently. In pregnant women, infections with Lm causes abortion and stillbirth. In infants and persons with a poor immune system it may lead to septicemia (blood poisoning) and meningitis. The disease is most often associated with consumption of foods such as soft cheese and processed meat products that are kept refrigerated for a long time, because Lm can grow at low temperatures.

Outbreaks of listeriosis have been reported from many countries, including Australia, Switzerland, France and the United States. Two recent outbreaks of Listeria monocytogenes in France in 2000 and in the USA in 1999 were caused by contaminated pork tongue and hot dogs respectively. Foodborne trematodes (worms) are increasing in South-east Asia and Latin America.

This is related to a combination of intensive aquaculture production in unsanitary conditions, and consumption of raw or lightly processed fresh water fish and fishery products.

Foodborne trematodes can cause acute liver disease, and may lead to liver cancer. It is estimated that 40 million people are affected worldwide. Bovine Spongiform Encephalopathy (BSE), is a fatal, transmittable, neurodegenerative disease of cattle. It was first discovered in the United Kingdom in 1985. The cause of the disease was traced to an agent in sheep, which contaminated recycled bovine carcasses used to make meat and bone meal additives for cattle feed.

Recycling of the BSE agent developed into a common source epidemic of more than 180,000 diseased animals in the UK alone. The agent affects the brain and spinal cord of cattle which produces sponge-like changes visible under a microscope. About 19 countries have reported BSE cases and the disease is no longer confined to the European Community.

A case of BSE has been reported in a cattle herd in Japan. In human populations, exposure to the BSE agent (probably in contaminated bovine-based food products) has been strongly linked to the appearance in 1996 of a new transmissible spongiform encephalopathy of humans called variant Creutzfeldt-Jakob Disease (vCJD). By January 2002,119 people developed vCJD, most from the UK but five cases have been reported from France.

Cancer and Environment

Cancer is caused by the uncontrolled growth and spread of abnormal cells that may affect almost any tissue of the body. Lung, colon, rectal and stomach cancer are among the five most common cancers in the world for both men and women. Among men, lung and stomach cancer are the most common cancers worldwide. For women, the most common cancers are breast and cervical cancer.

In India, oral and pharangeal cancers form the most common type of cancer which are related to tobacco chewing. More than 10 million people are diagnosed with cancer in the world every year. It is estimated that there will be 15 million new cases every year by 2020.Cancer causes 6 million deaths every year–or 12% of deaths worldwide. The causes of several cancers are known. Thus prevention of at least one-third of all cancers is possible.

Cancer is preventable by stopping smoking, providing healthy food and avoiding exposure to cancer-causing agents (carcinogens). Early detection and effective treatment is possible for a further one-third of cases. Most of the common cancers are curable by a combination of surgery, chemotherapy (drugs) or radiotherapy (X-rays). The chance of cure increases if cancer is detected early.

Cancer control is based on the prevention and control of cancer by:

- Promotion and strengthening of comprehensive national cancer control programs.
- Building international networks and partnerships for cancer control.
- Promotion of organized, evidence-based interventions for early detection of cervical and breast cancer.
- Development of guidelines on disease and programmes management.
- Advocacy for a rational approach to effective treatments for potentially curable tumours.
- Support for low-cost approaches to respond to global needs for pain relief and palliative care.

Prevention of Cancer

Tobacco smoking is the single largest preventable cause of cancer in the world. It causes 80 to 90% of all lung cancer deaths. Another 30% of all cancer deaths, especially in developing countries include deaths from cancer of the oral cavity, larynx, oesophagus and stomach which are related to tobacco chewing. Preventive measures include bans on tobacco advertising and sponsorship, increased tax on tobacco products, and educational programs which are undertaken to reduce tobacco consumption. Dietary modification is an important approach to cancer control.

Overweight individuals and obesity are known to be associated with cancer of the oesophagus, colon, rectum, breast, uterus and kidney. Fruit and vegetables may have a protective effect against many cancers. Excess consumption of red and preserved meat may be associated with an increased risk of colorectal cancer. Infectious agents are linked with 22% of cancer deaths in developing countries and 6% in industrialized countries.

Viral hepatitis B and C cause cancer of the liver. Human papilloma virus infection causes cancer of the cervix. The bacterium Helicobacter pylori increases the risk of stomach cancer. In some countries the parasitic infection schistosomiasis increases the risk of bladder cancer. Liver fluke increases the risk of cancer of the bile ducts. Preventive measures include vaccination and prevention of infection. Excessive solar ultraviolet radiation increases the risk of all types of cancer of the skin.

Avoiding excessive exposure to the sun, use of sunscreens and protective clothing are effective preventive measures. Asbestos is known to cause lung cancer. Aniline dyes have been linked to bladder cancer. Benzene can lead to leukaemia (blood cancer). The prevention of certain occupational and environmental exposure to several chemicals is an important element in preventing cancer.

IIUMAN RIGHTS

Several environmental issues are closely linked to human rights. These include the equitable distribution of environmental resources, the utilisation of resources and Intellectual Property Rights (IPRs), conflicts between people and wildlife especially around PAs, resettlement issues around development projects such as dams and mines, and access to health to prevent environment related diseases.

Equity

One of the primary concerns in environmental issues is how wealth, resources and energy must be distributed in a community. We can think of the global community, regional community issues, national concerns and those related to a family or at the individual level. While economic disparities remain a fact of life, we as citizens of a community must appreciate that a widening gap between the rich and the poor, between men and women, or between the present and future generations must be minimised if social justice is to be achieved.

Today the difference between the economically developed world and the developing countries is unacceptably high. The access to a better lifestyle for men as against women is inherent in many cultures. Last but not the least, we in the present generation cannot greedily use up all our resources leaving future generations increasingly impoverished. Rights to land, water, food, housing are all a part of our environment that we all share. However, while some live unsustainable lifestyles with consumption patterns that the resource base cannot support, many others live well below the poverty line. Even in a developing country such as ours, there are enormous economic inequalities.

This requires an ethic in which an equitable distribution becomes a part of everyone's thinking. The people who live in the countries of the North and the rich from the countries in the South will have to take steps to reduce their resource use and the waste they generate. Both the better off sectors of society and the less fortunate need to develop their own strategies of sustainable living and communities at each level must bring about more equitable patterns of wealth. The right to the use of natural resources that the environment holds is an essential component of human rights. It is related to disparities in the amount of resources available to different sectors of society.

People who live in wilderness communities are referred to as ecosystem people. They collect food, fuelwood, and nonwood products, fish in aquatic ecosystems, or hunt for food in forests and grasslands. When landuse patterns change from natural ecosystems to more intensively used farmland and pastureland the rights of these indigenous people are usually sacrificed. Take the case of subsidies given to the pulp and paper industry for bamboo which makes it several times cheaper for the industry than for a rural individual who uses it to build his home.

This infringes on the human right to collect resources they have traditionally used free of cost. Another issue is the rights of small traditional fishermen who have to contend against mechanised trawlers that impoverish their catch and overharvest fish in the marine environment. These people's right to a livelihood conflicts with the powerful economic interests of large- scale organised fisheries.

There are serious conflicts between the rights of rural communities for even basic resources such as water, and industrial development which requires large amounts of water for sustaining its productivity. The right to land or common property resources of tribal people is infringed upon by large development projects such as dams, mining and Protected Areas.

Movements to protect the rights of indigenous peoples are growing worldwide.

Reversing actions that have already been taken decades ago is a complex problem that has no simple solutions. In many cases a just tradeoff is at best achieved through careful and sensitively managed negotiations. This needs a deep appreciation of local environmental concerns as well as a sensitivity to the rights of local people.

Nutrition, health and Human Rights

There are links between environment, nutrition and health which must be seen from a humanrights perspective. Proper nutrition and health are fundamental human rights. The right to life is a Fundamental Right in our constitution. As a deteriorating environment shortens life spans, this in effect has an impact on our fundamental constitutional right. Nutrition affects and defines the health status of all people, rich and poor.

It is linked to the way we grow, develop, work, play, resist infection and reach our aspirations as individuals, communities and societies. Malnutrition makes people more vulnerable to disease and premature death. Poverty is a major cause as well as a consequence of ill-health. Poverty, hunger, malnutrition and poorly managed environments together affect health and weaken the socioeconomic development of a country.

Nearly 30% of humanity, especially those in developing countries– infants, children, adolescents, adults, and older persons are affected by this problem. A human rights approach is needed to appreciate and support millions of people left behind in the 20th century's health revolution. We must ensure that our environmental values and our vision are linked to human rights and create laws to support those that need a better environment, better health and a better lifestyle.

Health and sustainable human development are equity issues. In our globalized 21st century, equity must begin at the bottom, hand in hand with a healthy environment, improved nutrition, and sustainable lifestyles. Putting first things first, we must also realise that resources allocated to preventing and eliminating disease will be effective only if the underlying causes such as malnutrition and environmental concerns, as well as their consequences, are successfully addressed.

Intellectual Property Rights and Community Biodiversity Registers

Traditional people, especially tribals living in forests, have used local plants and animals for generations. This storehouse of knowledge leads to many new 'discoveries' for modern pharmaceutical products. The revenue generated from such 'finds' goes to the pharmaceutical industry that has done the research and patented the product. This leaves the original tribal user with nothing while the industry could earn billions of rupees.

To protect the rights of indigenous people who have used these products, a possible tool is to create a Community Biodiversity Register of local products and their uses so that its exploitation by the pharmaceutical industry would have to pay a royalty to the local community. This however has still not been generally accepted. Mechanisms have to be worked out so that the local traditional users rights are protected.

Traditional Medicine: Traditional medicine refers to health practices, approaches, knowledge and beliefs that incorporate plant, animal and mineral based medicines, frequently of local or regional origin. It may be linked to spiritual therapies, manual techniques and exercises. These may be used singly or in combination to treat, diagnose and prevent illnesses or maintain well- being. Traditional medicine is often handed down through the generations or may be known to a special caste or tribal group.

Traditional medicine has maintained its popularity in all regions of the developing world and its use is rapidly spreading in industrialized countries. In India, some of our primary health care needs are taken care of entirely by traditional medicine, while in Africa, up to 80% of the population uses it for primary health care. In industrialized countries, adaptations of traditional medicine are termed "Complementary" or "Alternative" Medicine (CAM). While there are advantages to traditional medicine as it is cheap and locally available, there are diseases which it cannot treat effectively.

This is a risk, as patients who use these alternative medicinal practices may rely on an ineffective measure. The consequences could be a serious delay in diagnosis and effective treatment of a treatable condition. There is a need to carefully research the claims of traditional practices to ensure that they are effective. In addition to patient safety issues, there is the risk that a growing herbal market and its great commercial benefit poses a threat to biodiversity through the over harvesting of the raw material for herbal medicines and other natural health care products.

This has been observed in the case of several Himalayan plants. If extraction from the wild is not controlled, this can lead to the extinction of endangered plant species and the destruction of natural habitats of several species. Another related issue is that at present, the requirements for protection provided under international standards for patent law and by most national conventional patent laws are inadequate to protect traditional knowledge and biodiversity.

There are tried and tested scientific methods and products that have their origins in different traditional medicinal methods. Twenty-five per cent of modern medicines are made from plants first used traditionally. Yoga is known to reduce asthma attacks. Traditional Medicine has been found to be effective against several infectious diseases. Over one-third of the population in developing countries lack access to essential allopathic medicines. The provision of safe and effective TM/CAM therapies could become a tool to increase access to health care.

VALUE EDUCATION

Value education in the context of our environment is expected to bring about a new sustainable way of life. Education both through formal and non-formal processes must thus address understanding environmental values, valuing nature and cultures, social justice, human heritage, equitable use of resources, managing common property resources and appreciating the cause of ecological degradation. Essentially, environmental values cannot be taught.

They are inculcated through a complex process of appreciating our environmental assets and experiencing the problems caused due to our destruction of our environment. The problems that are created by technology and economic growth are a result of our improper thinking on what 'development' means. Since we still put a high value only on economic growth, we have no concern for aspects such as sustainability or equitable use of resources. This mindset must change before concepts such as sustainable development can be acted upon.

Unsustainable development is a part of economic growth of the powerful while it makes the poor poorer. Consumerism is one aspect of this process favoured by the rich. As consumption of resources has till recently been an index of development, consumerism has thrived. It is only recently that the world has come to realise that there are other more important environmental values that are essential to bring about a better way of life.

Values in environment education must bring in several new concepts. Why and how can we use less resources and energy? Why do we need to keep our surroundings clean? Why should we use less fertilizers and pesticides in farms?

Why is it important for us to save water and keep our water sources clean? Or separate our garbage into degradable and non-degradable types before disposal? All these issues are linked to the quality of human life and go beyond simple economic growth. They deal with a love and respect for nature. These are the values that will bring about a better humanity, one in which we can live healthy, productive and happy lives in harmony with nature.

Environmental Values

Every human being has a great variety of feelings for different aspects of his or her surroundings. The Western, modern approach values the resources of Nature for their utilitarian importance alone. However true environmental values go beyond valuing a river for its water, a forest for its timber and non- wood forest products, or the sea for its fish.

Environmental values are inherent in feelings that bring about a sensitivity for preserving nature as a whole. This is a more spiritual, Eastern traditional value. There are several writings and sayings in Indian thought that support the concept of the oneness of all creation, of respecting and valuing all the different components of Nature. Our environmental values must translate to pro conservation actions in all our day to day activities.

Most of our actions have adverse environmental impacts unless we consciously avoid them. The sentiment that attempts to reverse these trends is enshrined in our environmental values. Values lead to a process of decision making which leads to action. For value education in relation to the environment, this process is learned through an understanding and appreciation of Nature's oneness and the importance of its conservation.

Humans have an inborn desire to explore Nature. Wanting to unravel its mysteries is a part of human nature. However, modern society and educational processes have invariably suppressed these innate sentiments. Once exposed to the wonders of the wilderness, people tend to bond closely to Nature.

They begin to appreciate its complexity and fragility and this awakens a new desire to want to protect our natural heritage. This feeling for Nature is a part of our Constitution, which strongly emphasises this value. Concepts of what constitutes right and wrong behaviour changes with time. Values are not constant.

It was once considered 'sport' to shoot animals. It was considered a royal, brave and much desirable activity to kill a tiger. In today's context, with wildlife reduced to a tiny fraction of what there was in the past, it is now looked down upon as a crime against biodiversty conservation. Thus the value system has been altered with time. Similarly with the large tracts of forest that existed in the past, cutting a few trees was not a significant criminal act.

Today this constitutes a major concern. We need a strong new environmental value system in which felling trees is considered unwise behaviour. With the small human numbers in the past, throwing away a little household degradable garbage could not have been considered wrong. But with enormous numbers of people throwing away large quantities of non-degradable waste, it is indeed extremely damaging to the environment and our value system must prevent this through a strong environmental value education system.

Appreciating the negative effects of our actions on the environment must become a part of our day to day thinking. Our current value system extols economic and technical progress as being what we need in our developing country.

Environmental Values Based on the Constitution of India

- *Article 48A*: "The state shall endeavour to protect and improve the environment and to safeguard the forests and wildlife in the country."
- *Article 51A (g)*: The constitution expects that each citizen of the country must "protect and improve the natural environment, including forests, lakes, rivers and wildlife, and to have compassion for all living creatures."

While we do need economic development, our value system must change to one that makes people everywhere support a sustainable form of development so that we do not have to bear the cost of environmental degradation. Environmental problems created by development are due neither to the need for economic development, nor to the technology that produces pollution, but rather to a lack of awareness of the consequences of unlimited and unrestrained anti-environmental behaviour.

Looked at in this way, it deals with concepts of what is appropriate behaviour in relation to our surroundings and to other species on Earth. How we live our lives in fact shapes our environment. This is what environmental values are about. Each action by an individual must be linked to its environmental consequences in his/her mind so that a value is created that leads to strengthening pro-environmental behaviour and preventing anti- environmental actions.

This cannot happen unless new educational processes are created that provide a meaning to what is taught at school and college level. Every small child while growing up asks questions like 'What does this mean?'. They want an explanation for things happening around them that can help them make decisions and through this process develop values. It is this innate curiosity that leads to a personalized set of values in later life.

Providing appropriate 'meanings' for such questions related to our own environment brings in a set of values that most people in society begin to accept as a norm. Thus pro environmental actions begin to move from the domain of individuals to that of a community. At the community level, this occurs only when a critical number of people become environmentally conscious so that they constitute a proenvironment lobby force that makes governments and other people accept good environmental behaviour as an important part of development. What professions require making value judgements that greatly influence our environment? Evidentally nearly every profession can and does influence our environment, but some do so more than others. Policy makers, administrators, landuse planners, media, architects, medical personnel, health care workers, agriculturalists, agricultural experts, irrigation planners, mining experts, foresters, forest planners, industrialists and, most importantly, teachers at school and college level, are all closely related to pro environmental outcomes.

Environmental values have linkages to varied environmental concerns. While we value resources that we use as food, water and other products, there are also environmental services that we must appreciate. These include Nature's mechanisms in cleaning up air by removing carbon dioxide and adding oxygen by plant life, recycling water through the water cycle of nature, maintaining climate regimes, etc. But there are other aesthetic, ethical values that are equally important aspects of our environment that we do not appreciate consciously.

While every species is of importance in the web of life, there are some which man has come to admire for their beauty alone. The tiger's magnificence, the whale and elephant's giant size, the intelligence of our cousins the primates, the graceful flight of a flock of cranes, are parts of nature that we cannot help but admire. The lush splendor of an evergreen forest, the great power of the ocean's waves, and the tranquility of the Himalayan mountains are things that each of us values even if we do not experience it ourselves. We value its being there on Earth for us.

This is called its 'existance value'. The list of wondrous aspects of Nature's intricate connections is indeed awe-inspiring. This is also a part of our environment that we must value for its own sake. This is the oneness of Nature. We must equally look at our environment beyond the wild sphere. There is incredible beauty in some man-modified landscapes, the coloured patterns of farmland or the greens of a tea or coffee plantation in the hills.

Urban gardens and open space are also valuable and thus must be of prime concern to urban planners. These green spaces act as not only the lungs of a city, but also provide much needed psychological support. The mental peace and relaxation provided by such areas needs to be valued, although it is difficult to put a price tag on these values.

Nevertheless, these centers of peace and tranquility give urban dwellers an opportunity to balance their highly man-modified environments with the splash of green of a garden space. Environmental values must also stress on the importance of preserving ancient structures. The characteristic architecture, sculpture, artworks and crafts of ancient cultures is an invaluable environmental asset.

It tells us where we have come from, where we are now, and perhaps where we should go. Architectural heritage goes beyond preserving old buildings, to conserving whole traditional landscapes in rural areas and streetscapes in urban settings. Unless we learn to value these landscapes, they will disappear and our heritage will be lost.

As environmentally conscious individuals we need to develop a sense of values that are linked with a better and more sustainable way of life for all people. There are several positive as well as negative aspects of behaviour that are linked to our environment. The positive feelings that support environment include a value for Nature, cultures, heritage, and equity. We also need to become more sensitive to aspects that have negative impacts on the environment. These include our attitude towards degradation of the environment, loss of species, pollution, poverty, corruption in environmental management, the rights of future generations and animal rights. Several great philosophers have thoughts that have been based on, or embedded, in pro environmental behaviour.

Mahatma Gandhi and Rabindranath Tagore are among the many internationally well-known scholars whose thought have included values that are related to environmental consciousness. We need to appreciate these values to bring about a better way of life on earth for all people and all living creatures.

Valuing Nature

The most fundamental environmental sentiment is to value Nature herself. Appreciating Her magnificence and treasuring life itself leads to positive feelings that are a manifestation of pro environmental consciousness. The one- ness of our lives with the rest of nature and a feeling that we are only a miniscule part of nature's complex web of life becomes apparent, when we begin to appreciate the wonders of nature's diversity. We must appreciate that we belong to a global community that includes another 1.8 million known living forms.

Nothing makes us more conscious of this wonderous aspect of our earth's diversity than a walk through the wilderness, feeling and exploring its beauty and experiencing its infinite variety. The tiny creatures that live complex lives and the towering trees are all a part of this phenomenon we call 'life'. Today, man does not even know if other complex forms of life exist outside our own solar system in distant space. We may be alone in space or may be accompanied by other, completely different, living forms. But for now we only know for sure that the Earth's life forms are unique.

We thus have a great responsibility to protect life in all its glorious forms and must therefore respect the wilderness with all its living creatures, where man's own hand has not created changes that have led to perturbing natural habitats. We need to develop a sense of values that lead us to protect what is left of the wilderness by creating effective National Parks and Wildlife Sanctuaries. However this cannot be done to the detriment of the millions of tribal or indigenous people who live in wilderness ecosystems. There are thus conflicting values that need to be balanced carefully. On the one hand we need to protect natural ecosystems, while on the other, we must protect the rights of local people. Yet apart from valuing the diversity of life itself, we must also learn to value and respect diverse human cultures. Many of the tribal cultures of our country are vanishing because those with more dominant and economically advanced ways of life do not respect their lifestyles, that are in fact closer to nature and frequently more sustainable. We believe that our modern technology-based lifestyles are the sole way for society to progress. Yet this is only a single dimension of life that is based on economic growth.

Deep Ecology

In the 1970s a new thinking on environmental concerns began to emerge, protecting nature and the wilderness for its own sake, which is now referred to as 'Deep Ecology'.

It was fostered by the thinking of Arne Naess, a Norwegian professor of Philosophy and a great believer in Gandhian thinking and Buddhism. It recognises the intrinsic value of all living beings and looks upon mankind as a small segment of a great living community of life forms.

It teaches that the wellbeing and flourishing of human and non-human life on Earth have value in themselves and that these values are independent of the usefulness of the non-human world for human purposes. While currently the environmental movement focuses on issues that are concerned with the management of the natural environment for the 'benefit' of man, Deep Ecology promotes an approach that is expected to bring about a more appropriate ecological balance on Earth and is akin to a spiritual approach to Nature. For example some environmentalists emphasise the need to preserve wilderness for its aesthetic and utilitarian functions.

Wilderness is being preserved today in PAs because it is scenic and serves the purpose of tourism for nature lovers, and has recreational and economic value. Other environmentalists stress that the goal is for protecting the useful ecological functions of the wilderness, its services and goods that we use. Deep Ecologists on the other hand stress that wilderness preservation is a means to achieve the conservation and protection of biological diversity. Thus it is not enough to protect bits of what is left of the wilderness but to make attempts to restore degraded areas to their former natural ecological state.

In a country such as India, with its enormous population coupled with poverty on the one hand and the need for economic industrial growth on the other, this will be extremely difficult to achieve. Another new approach is that of 'Gaia', the hypothesis that the Earth is itself like one giant form of throbbing life consisting of all the unquantifiable numbers of individuals of its millions of known and unknown species.

Valuing Cultures

Every culture has a right to exist. Tribal people are frequently most closely linked with Nature and we have no right to foist on them our own modern way of life. The dilemma is how to provide them with modern health care and education that gives them an opportunity to achieve a better economic status without disrupting their culture and way of life. This will happen only if we value their culture and respect their way of life.

Social Justice

As the divide widens between those people who have access to resources and wealth, and those who live near or below the poverty line, it is the duty of those who are better off to protect the rights of the poor who do not have the means to fight for their rights. If this is not respected the poor will eventually rebel, anarchy and terrorism will spread and the people who are impoverished will eventually form a desperate seething revolution to better their own lot.

The developing world would face a crisis earlier than the developed countries unless the rights of poor people that are fundamental to life are protected.

Modern civilization is a homogenous culture, based until recently on a belief that modern science holds the answer to everything. We are now beginning to appreciate that many ancient and even present day sequestrated cultures have a wisdom and knowledge of their own environments that is based on a deep sense of respect for nature.

Tribal cultures have over many generations used indigenous medicines which are proving to be effective against diseases. They have produced unique art forms such as painting, sculpture, and crafts that are beautiful and can enrich living experiences for everyone.

They have their own poetry, songs, dance and drama-all art forms that are unfortunately being rapidly lost as we introduce a different set of modern values to them through television and other mass media.

The world will be culturally impoverished if we allow these indigenous people to loose their traditional knowledge which includes sustainable use of water, land and resources with a low impact on biodiversity.

They will soon lose the beauty within their homes that is based on the things they make from Nature.

The art of the potter will be lost forever to the indestructible plastic pot. The bamboo basket weaver who makes a thing of beauty that is so user friendly and aesthetically appealing, will give place to yet another plastic box. Much that is beautiful and hand-crafted will disappear if we do not value these diverse aspects of human cultures.

Human Heritage

The earth itself is a heritage left to us by our ancestors for not only our own use but for the generations to come. There is much that is beautiful on our Earth-the undisturbed wilderness, a traditional rural landscape, the architecture of a traditional village or town, and the value of a historical monument or place of worship. These are all part of human heritage. Heritage preservation is now a growing environmental concern because much of this heritage has been undervalued during the last several decades and is vanishing at an astonishing pace.

While we admire and value the Ajanta and Ellora Caves, the temples of the 10th to 15th centuries that led to different and diverse styles of architecture and sculpture, the Moghul styles that led to structures such as the Taj Mahal, or the unique environmentally-friendly Colonial buildings, we have done little to actively preserve them. As environmentally conscious individuals we need to lobby for the protection of the wilderness and our glorious architectural heritage.

Equitable use of Resources

An unfair distribution of wealth and resources, based on a world that is essentially only for the rich, will bring about a disaster of unprecedented proportions. Equitable use of resources is now seen as an essential aspect of human well being and must become a shared point of view among all socially and environmentally conscious individuals.

This includes an appreciation of the fact that economically advanced countries and the rich in even poor nations consume resources at much greater levels than the much larger poorer sectors of humanity in the developing world. In spite of the great number of people in the more populous developing countries, the smaller number of people in developed countries use more resources and energy than those in the developing world. This is equally true of the small number of rich people in poor countries whose per capita use of energy and resources, and the generation of waste based on the one time use of disposable products, leads to great pressures on the environment. The poor while polluting the environment have no way to prevent it.

The rich damage the environment through a carelessness that proves only that they have no appreciation for environmental safety. As we begin to appreciate that we need more sustainable lifestyles we also begin to realise that this cannot be brought about without a more equitable use of resources.

Common Property Resources

Our environment has a major component that does not belong to individuals. There are several commonly owned resources that all of us use as a community. The water that nature recycles, the air that we all breathe, the forests and grasslands which maintain our climate and soil, are all common property resources. When Government took over the control of community forests in British times, the local people who until then had controlled their use through a set of norms that were based on equitable use, began to overexploit resources on which they now had no personal stake.

Bringing back such traditional management systems is extremely difficult. However, in the recent past managing local forests through village level forest protection committees has shown that if people know that they can benefit from the forests, they will begin to protect them. This essentially means sharing the power to control forests between the Forest Department and local people.

Ecological Degradation

In many situations valuable ecological assets are turned into serious environmental problems. This is because we as a society do not strongly resist forces that bring about ecological degradation. These consist of sectors of society that use a 'get-rich-quick' approach to development. While ecological degradation has frequently been blamed on the needs of fuelwood and fodder of growing numbers of rural people, the rich, urbanized, industrial sector is responsible for greater ecological damage.

Changes in landuse from natural ecosystems to more intensive utilization such as turning forests into monoculture forestry plantations, or tea and coffee estates, or marginal lands into intensive agricultural patterns such as sugarcane fields or changes into urban or industrial land carry an ecological price. Wetlands, for example, provide usable resources and a variety of services not easily valued in economic terms, and when destroyed to provide additional farmland, in many cases produce lower returns. A natural forest provides valuable non-wood forest products whose economic returns far outweigh that provided by felling the forest for timber. These values must form a part of a new conservation ethic. We cannot permit unsustainable development to run onwards at a pace in which our lives will be overtaken by a development strategy that must eventually fail as Earth's resources are consumed and ecosystems rendered irreparable.

HIV/AIDS

The Human Immunodeficiency Virus (HIV) causes Acquired Immunodeficiency Syndrome (AIDS) through contact with tissue fluids of infected individuals, especially through sexual contact. As it reduces an individual's resistance to disease, it causes infected individuals to suffer from a large number of environment related diseases and reduces the ability of infected individuals to go about their normal lives.

It affects their income generation and/or their ability to utilise natural resources. As more and more people are affected, this disease will also have impacts on our natural resource base, as utilisation patterns change to unsustainable levels. The inability of these patients to have the strength to access natural resources also affects the outcome of the disease process, as their overall health and well being is likely to worsen the course of the disease when their nutritional status suffers. In sub Saharan Africa where the infection has become highly prevalent, it is leading to great suffering and worsening poverty.

The capacity of these patients to work for their usual sources of income generation is lost. An increasing proportion of the poor are affected. It is evident that it is going to be increasingly difficult to manage environments sustainably, as natural resources on which the poor debilitated patients depend continue to be degraded. Incomes lost due to the stigma of HIV/AIDS must be met by the sufferers by overexploiting their resource base. People affected by the disease inevitably try to get whatever they can from their natural resource base as they are not in any position to think of the long-term future.

In Africa, this has led to degradation of the ecosystem and an increase of pressures from other impacts such as overuse of medicinal plants and poaching for wildlife. In South Africa, for example, people have a mistaken belief that turtle eggs can cure HIV/AIDS, thus leading to the eggs being over harvested. As males die of the disease, work on agricultural land has to be taken over by already overworked women and their children, affecting land management and productivity. Providing balanced diets and nutritional support for these poverty stricken patients can be partially addressed by better natural resource management such as afforestation, access to clean water and wholesome food. HIV/AIDS seriously affects the patient's working environment. It creates an incorrect fear in the minds of co-workers. It must be clearly understood that AIDS is not spread by casual contact during work. Patients have a right to continue to work as before along with unaffected individuals. As patients are unable to continue their original hard labour related work, it is essential that alternative sources of work must be created for them. Educators and extention information, in the formal and non-formal educational sectors, must address the issues related to the linkages between natural resource management and this disease, as well as the need to remove the social stigma attached to it.

HIV/AIDS has a serious impact on the socioeconomic fabric of society. By 2002, India had an estimated 3.97 million infected individuals. There is a great need to organise AIDS education on prevention and management of the disease. This needs to be done through the formal educational sector and by using non-formal methods. Education is also important to reduce the stigma and discrimination against these patients. In India, women who are not socially empowered are at a great disadvantage as they are powerless to demand safe sex from their partners. Women also have an added burden of caring for HIV infected husbands. This produces enormous economic stresses on their family. HIV in India is rapidly moving from a primarily urban sector disease to rural communities. Research in Nepal has shown a linkage between rural poverty, deforestation and a shift of population to urban areas resulting in a rising number of AIDS patients. Prior to 1992, it was mainly seen in males who migrated to urban centers. In more recent times, a growing number of women are moving to Indian cities as sex workers. Women engaged in prostitution find it difficult to make partners take protective measures, such as the use of condoms that provide safe sex. A large proportion became victims of the disease. Blood transfusion from an infected person can also lead to HIV/AIDS in the recipient, as well as drug abuse by sharing needles with an infected person. In sexually transmitted AIDS, the use of condoms during intercourse is a key to preventing the disease. Behavioural change, where the number of individuals who have multiple partners, towards strictly single partners, reduces the risk of HIV/AIDS and thus reduces incidence of the disease in society. However, the most important measure to prevent AIDS is the proper use of condoms that form a barrier to the spread of the virus during intercourse.

WOMAN AND CHILD WELFARE

There are several environmental factors that are closely linked to the welfare of women and children. Each year, close to eleven million children worldwide are estimated to have died from the effects of disease and inadequate nutrition. Most of these deaths are in the developing world. In some countries, more than one in five children die before they are 5 years old. Seven out of 10 of childhood deaths in developing countries can be attributed to five main causes, or a combination of them. These are pneumonia, diarrhoea, measles, malaria and malnutrition. Around the world, three out of every four children suffer from at least one of these conditions.

Table. The Diagnosis of Common Childhood Disease Conditions

Presenting complaint	Possible cause or associated condition
Cough and/or fast breathing	Pneumonia Severe anaemia P. falciparum malaria
Lethargy or unconsciousness	Cerebral malaria Meningitis Severe dehydration Very severe pneumonia
Measles rash	Pneumonia Diarrhoea Ear infection
Very sick" young infant"	Pneumonia Meningitis Sepsis

RespiratoryConditions

Most respiratory diseases are caused by or are worsened by polluted air.

Crowded ill-ventilated homes and living in smokey households with open fires can trigger respiratory conditions especially in children. Pneumonia: Acute respiratory infections (ARI), most frequently pneumonia, is a major cause of death in children under five, killing over two million children annually.

Upto 40% of children seen in health centers suffer from respiratory conditions and many deaths attributed to other causes are, in fact, "hidden" ARI deaths. Children may die very quickly from the infection and thus need treatment urgently. Most patients of pneumonia can be treated with oral antibiotics. Correct management could save over 1 million lives per year globally.

Gastro Intestinal Conditions

Contaminated water and food causes widespread ill health especially in children. Diarrhoea: Diarrhoea is caused by a wide variety of infections. Urgent diagnosis and treatment of diarrhoea is a priority for saving a child's life. Treating malnutrition that often accompanies diarrhoea can further reduce mortality. Increasing vigilance to detect other diseases that can occur concurrently with diarrhoea, such as measles or malaria, is an important measure.

Two million children die each year in developing countries from diarrhoeal diseases, the second most serious killer of children under five worldwide. In most cases diarrhoea is preventable and children can be saved by early treatment. Correct management of diarrhoea could save the lives of up to 90% of children who currently die by promoting rapid and effective treatment through standardised management, including antibiotics and simple measures such as oral rehydration using clean boiled water with salt and sugar.

In severe cases intravenous fluids must be started. Improved hygiene and management of the home and surroundings is the most important preventative measure, as well as improved nutrition. Increased breastfeeding and measles vaccination have also been observed to have reduced the number of cases of diarrhoea.

Measles

Measles is a rash that appears with fever and bodyache in children and is caused by a virus. It infects over 40 million children and kills over 800,000 children under the age of five. Prevention includes wider immunization coverage, rapid referral of serious cases, prompt recognition of conditions that occur in association with measles, and improved nutrition, including breastfeeding, and vitamin A supplementation. Measles is prevented by a vaccine. Young children with measles often develop other diseases such as acute respiratory infections, diarrhoea and malnutrition that are all linked to poor environmental conditions in their surroundings. Children who survive an attack of measles are more vulnerable to other dangerous infections for several months. Effective prevention and treatment could save 700,000 lives per year.

Malaria

This condition is closely linked to pooling and stagnation of water in tropical environments. Malaria is a widespread tropical disease which is caused by a parasite transmitted to humans by mosquitoes. It has proved difficult to control because mosquitoes have become resistant to insecticides used against them and because the parasite has developed resistance in some areas to the cheap and effective drugs that used to provide good protection in the past.

However, alternative newer drug therapies have been developed for use in areas where resistant parasites are found. In India the disease was nearly wiped out a few decades ago but has now re-emerged in many parts of the country. Correct management could save 500,000 lives per year. Approximately 700,000 children die of malaria globally each year, most of them in sub-Saharan Africa. Young children are particularly vulnerable because they have not developed the partial immunity that results from surviving repeated infections. Deaths from malaria can be reduced by several measures, including encouraging parents to seek prompt care, accurate assessment of the condition of the child, prompt treatment with appropriate anti-malarial drugs, recognition and treatment of other co-existing conditions, such as malnutrition and anaemia, and prevention by using mosquito-proof bednets. Because fever may be the only sign of malaria, it can be difficult to distinguish it from other potentially lifethreatening conditions.

Poverty-environment-malnutrition

There is a close association between poverty, a degraded environment, and malnutrition. This is further aggravated by a lack of awareness on how children become malnourished. Malnutrition: Although malnutrition is rarely listed as the direct cause of death, it contributes to about half of all childhood deaths. Lack of access to food, poor feeding practices and infection, or a combination of the two, are major factors in mortality.

Infection, particularly frequent or persistent diarrhoea, pneumonia, measles and malaria, undermines nutritional status. Poor feeding practices-inadequate breastfeeding, providing the wrong foods, giving food in insufficient quantities, contribute to malnutrition. Malnourished children are more vulnerable to disease. Promoting breastfeeding, improving feeding practices, and providing micronutrient supplements routinely for children who need them are measures that reduce mortality. The nutritional status and feeding practices of every child under two years of age, and those with a low weight for their age must be intensively managed. Counseling of parents on the correct foods for each age group and helping them to overcome various feeding problems is an essential health care measure. Children between 6 months and 2 years of age are at increased risk of malnutrition when there is a transition between breastfeeding and sharing fully in the family diet.

Changing family habits and the kinds of food offered to children is an important measure. Talking to mothers individually about home care and their child's feeding, with relatively simple changes to better feeding practices, such as helping them to eat rather than leaving them to fend for themselves, can ensure that a child gets enough to eat. A minor increase in breastfeeding could prevent up to 10% of all deaths of children under five: When mothers breastfeed exclusively during at least the first four months and, if possible, six months of life, there is a decrease in episodes of diarrhoea and, to a lesser extent, respiratory infections.

Even small amounts of water-based drinks decreases breastmilk intake and lead to lowered weight gain. This increases the risk of diarrhoea. Continuing to breastfeed up to two years of age, in addition to giving complementary foods, maintains good nutritional status and helps prevent diarrhoea.

Encouraging maximum support to mothers to establish optimal breastfeeding from birth, equipping health workers with counseling skills, and providing individual counseling and support for breastfeeding mothers are measures that reduce malnutrition.

Mothers often give their babies other food and fluids before six months because they doubt their breastmilk supply is adequate. A one-on-one counseling with mothers on breastfeeding techniques and its benefits helps reduce incidence of malnutrition. There are strong connections between the status of the environment and the welfare of women and children in India. Women, especially in lower income group families, both in the rural and urban sector, work longer hours than men.

Their work pattern differs and is more prone to health hazards. The daily collection of water, fuelwood and fodder is an arduous task for rural women. In urban areas, where lower economic group women live in crowded smoke filled shantys in unhygenic slums, they spend long hours indoors, which is a cause of respiratory diseases.

In urban centers, a number of women eke out a living by garbage picking. They separate plastics, metal and other recyclable material from the waste produced by the more affluent groups of society. During this process, they can get several infections. Thus they are providing an environmental service of great value, but earn a pittance from this work.

Women are often the last to get enough nutrition as their role in traditional society is to cook the family meal and feed their husband and children. This leads to malnutrition and anemia due to inadequate nutrition. The sorry plight of women includes the fact that the girl child is given less attention and educational facilities as compared to boys in India.

Case Study

Karnataka's GIS scheme, Bhoomi, has revolutionized the way farmers access their land records. Farmers can now get a copy of the records of rights, tenancy and crops from a computerized information kiosk without harassment and bribes. Karnataka has computerized 20 million records of land ownership of 6.7 million farmers in the State. they are unable to compete with men in later life. This social-environmental divide is a major concern that needs to be corrected throughout the country.

ROLE OF INFORMATION TECHNOLOGY IN ENVIRONMENT AND HUMAN HEALTH

The understanding of environmental concerns and issues related to human health has exploded during the last few years due to the sudden growth of Information Technology. The computer age has turned the world around due to the incredible rapidity with which IT spreads knowledge. IT can do several tasks extremely rapidly, accurately and spread the information through the world's networks of millions of computer systems.

A few examples of the use of computer technology that aid environmental studies include software such as using Geographical Information Systems (GIS). GIS is a tool to map landuse patterns and document change by studying digitized toposheets and/or satellite imagery. Once this is done, an expert can ask a variety of questions which the software can answer by producing maps which helps in landuse planning. The Internet with its thousands of web sites has made it extremely simple to get the appropriate environmental information for any study or environmental management planning. This not only assists scientists and students but is a powerful tool to help increase public awareness about environmental issues. Specialised software can analyse data for epidemiological studies, population dynamics and a variety of key environmental concerns.

The relationship between the environment and health has been established due to the growing utilisation of computer technology. This looks at infection rates, morbidity or mortality and the etiology (causative factors) of a disease. As knowledge expands, computers will become increasingly efficient. They will be faster, have greater memories and even perhaps begin to think for themselves.

❐

5

Population Change, and the Rural Environment

Rural environment represents the framework of regulations, institutions, and practices in villages defining parameters for the sustainable use of environmental resources while ensuring security of livelihood and a reasonable quality of life. While the scope of environmental infrastructure is often narrowed down to the provision of suitable water supply, sewerage, and sanitation systems, it has within its purview:

- Acquisition, protection, and maintenance of open spaces,
- Clean up and restoration of degraded lands,
- Integration of existing wildlife or habitat resources,
- Sustainable approaches to controlling flooding and drainage,
- Developing river corridors and coastal areas,
- Forest management.

Rejuvenation of natural resources through activation of watersheds, renewal of wastelands along with enhancement of farm productivity, is acomponent of environmental infrastructure that is attaining increasingimportance as expanding anthropogenic activity stresses natural resources beyond their natural regenertion capability. We focus here on naturalresources, common properties, and rejuvenation of rural environment. Here we take stock of the rural environment and propose institutional mechanisms to keep the juggernaut of socioeconomic development rolling without impediments. We present a snapshot of the current rural environment demonstrating the phenomena through which irreversible degradation of the environment has resulted.

We examine the veracity of the widely held position that poor social and economic conditions of villagers compel them to overly exploit the environment, leading to a vicious circle of degradation of natural resources perpetuating poverty. Possible policy and technological solutions to pre-empt environmental degradation, restore, and rejuvenate the degraded environment are discussed further on in the chapter.

STATUS OF THE RURAL ENVIRONMENT

The ecosystem within which all rural activities are conducted encompasses the air we breathe, the water bodies surrounding us, and the land we walk on. Unfettered human activity can compromise the ability of the environment to support our kind, a condition usually referred to as environmental degradation and detrimental to our future.

India supports approximately 16 per cent of the world population and 20 per cent of its livestock on 2.5 per cent of its geographical area, making its environment a highly stressed and vulnerable system. The pressure on land has led to soil erosion, water logging, salinity, nutrient depletion, lowering of the groundwater table, and soil pollution—largely a consequence of thoughtless human intervention.

The extent of land degradation, the loss in capacity of our major water reservoirs and the decline in water level in wells in the past few years is alarming. Soil erosion from overgrazing, and intensive cultivation and soil degradation from excessive use of agricultural chemicals, have wide-ranging implications. Agricultural activities that cause land degradation include shifting cultivation without adequate fallow periods, absence of soil conservation measures, cultivation of fragile lands, unbalanced fertilizer use, faulty planning or management of irrigation.

Improper agricultural practices are usually observed under constraints of saturation of good lands and population pressure leading to cultivation of 'too shallow' or 'too steep' soils and ploughing of fallow land before it has recovered its fertility. Overgrazing and over-extraction of green fodder lead to forest degradation through decreased vegetative regeneration, compaction of soil, and reduced infiltration and vulnerability to erosion. Annual environmental costs for India in 1995 were estimated at US$ 9.7 billion, of which surface water pollution, land degradation and deforestation contributed 84 per cent. Deforestation with shifting agriculture, over-exploitation for fuel wood and timber collection, and mining activities are also causes of serious concern. Deforestation causes degradation when the land is steeply sloping, or has shallow or easily erode-able soils, and when clearance is not followed by good management.

Over-cutting of vegetation to obtain timber, fuel wood and other products is frequent in semi-arid environments, where fuel wood shortages are severe. Overgrazing causes a decrease in vegetation cover which is a primary cause of erosion. According to the data provided by the National Remote Sensing Agency and Forest Survey of India based on satellite imagery, 80 Mha of 142 Mha of land under cultivation is substantially degraded and about 40 Mha of 75 Mha of land under the forest department has a canopy density cover less than 40 per cent. Perhaps the most widely recognized environmental problem is the pollution of water resources by industrial discharge, household waste, sewage, and agricultural chemicals. Water scarcity induced by mounting population density and growing economic activity in the face of fixed resources, depleting water tables, and silting of reservoirs has led to rapid decline in the quality of life in rural India. Existing irrigated areas are displaying serious water-stressed situations, as both reservoirs and groundwater sources continue to get depleted.

Consequently, the agricultural output from irrigated areas also seems to be more vulnerable to weather shocks than earlier. The problem is compounded by the fact that provision of cheap power encourages farmers to use excessive water. While this problem is widely acknowledged, a holistic policy framework to address the problem effectively is missing. India's biodiversity is gradually narrowing. Maintaining viable populations of species—whether plant or animal—is crucial in biodiversity conservation requiring the protection of important ecosystems, habitats, and the ecological processes of which they are a part. The loss of biodiversity in shrinking forests as well as in threatened marine and wetland ecosystems has strong adverse impacts in store.

ECOLOGICAL FOOTPRINT

Ecological footprint analysis is an accounting tool that estimates the resource consumption and waste assimilation requirement of a defined human population or economy, in terms of a corresponding land area. Such an analysis encompasses many human needs and aggregates them into a single figure, expressed in global hectares per capita per year, which can be used to shape environmental policies. Because of the complexities involved in determining an eco-footprint, very little analysis has been undertaken at the rural scale globally, although country eco-footprints are estimated and published annually by the World Wildlife Fund. The global eco-footprint, based on 2001 data was 13.5 billion global hectares, which exceeds the global bio-capacity by 21 per cent, or 0.4 global hectares per person.

Although India's eco-footprint over the last forty years is almost constant at around 1 global hectare per person (compared to the global average of 1.8), available bio-capacity per person has fallen as its population has almost doubled. The consumption patterns of the large middle class, residing mainly in urban areas will shape its eco-footprint in the years to come and it is here that the rural areas have a role to play in holding the footprint down to reasonable levels by engaging in sustainable farming practices. It must also be noted that the footprint is a very human-centric concept, talking about hectares of land to be set aside for humans without any consideration for other species.

IMPACT ON HUMAN HEALTH

Globally, among the biggest dangers from farming is the continuous exposure to and the unsafe use of chemicals necessary for agriculture. In India, however, the danger to human health from such environment and pollution related causes are not given their due importance as accidents from farm machinery, with a fatality rate of 22 per 100,000 farmers.

Fatality apart, chronic exposure to air and waterborne chemicals can have adverse health effects, which sometimes, can be difficult to measure because of problems in isolating individual chemical effects.

While certain cause and effect relationships are not easy to identify, cumulative effects are likely to be most critical. Cancer risk could be high from nitrate, metals, as well as pesticides; other problems like adverse hormonal functions, liver damage could also take place, as summarized in Table. Moreover, toxic chemicals and pesticides in air, water, and earth enter body tissues and breast milk, through which they are passed on to infants.

For example, the impact of spraying endrin and endosulfan on cashew plantations in the 1970s to combat pests in the Kasargod district of Kerala has been so persistent that in one village alone, 156 cases of health disorders were recorded from 123 households between 1990 and 2001. These included 49 cancer cases and 23 mental retardation cases among several others. Moreover, endosulfan residues, several times the minimum permissible limit, were found in water, milk, vegetable, and fish, putting the village communities at risk even three decades after spraying of the pesticideS

Table. Impact of Agro-Chemicals on Human Health

Agro chemicals	Maximum Contamination Levels	Health Effects	
		Established	Potential
Nitrate	10mg/NO3-N	Methaglominemia	Nitrate conversion to nitrite and N-Nitroso compounds affecting thyroid, endocrine functions
Metals (Cd, As, Cr, Zn, Cu) Pesticides (chlorinated hydrocarbons like DDT, organophosphate like Malathion, Carbamata like Carbaryl)	0.01 to 0.05 mg/L	Impaired kidney functions, skin disorder, tumour Affects nervous system and enzymes reproduction, affects and muscles	Cancer risk cancer risk

The State of Kerala's Environment

The 'Kerala model' of development has achieved remarkable milestones in the fields of health, literacy, housing, and so on, and earned worldwide praise for the quality of life of its people. The state is now poised for a quantum leap with greater involvement of the common people through what is now called as the 'people's campaign'. It has identified sectors such as tourism, aquaculture, power production, and software production as priority areas for people's participation. Kerala has long been described as 'God's Own Country' in tourism circles.

However, the state of Kerala's forests and rivers is alarming. The total forest area of the state has come down from 11,241.97 ha in 1994 to 11,236.06 ha in 1996: this means 6 ha of forest has vanished in two years. The forest cover in the Periyar catchment area alone is reported to have dwindled by two-thirds over the last hundred years. Deforestation and encroachment by settlers, with large-scale illegal felling of trees by private estate owners and others, have received frequent media attention but hardly enough policy attention. Soil erosion has resulted along with change in climate patterns, as well as movements of wild animals, leading to increasing conflict between man and animal, and disappearance of rare and endangered flora and fauna. Destruction of the forest and tree cover in Nelliampathy hills have resulted in growing incidence of poaching. Ground-clearing for plantations have led to the destruction of rare and endangered orchid species. Kerala's rivers are assailed by pollution, sedimentation, sand mining, and constriction of flows. Garbage and animal wastes from slaughterhouses as well as sewage water are released directly into rivers. A recent survey of twelve major rivers of the state, including the Periyar, the Bharathapuzha, the Pamba, the Chaliyar, and the Valapattanam, highlighted presence of high levels of coliform bacteria. The problem was reported to be the most serious in the Pamba with pilgrims causing most of the damage.

Soil erosion, caused by deforestation and various agricultural, mining, and construction activity, has led to large-scale siltation of the water bodies. The problem is particularly acute in the case of reservoirs. It has reduced their water-holding capacity and affected the utility of the dams. Siltation at the mouths of rivers causes, besides navigation bottlenecks, flooding and damage to the environment. Siltation in the Vembanad Lake seems to have reached a critical stage.

It threatens to turn the lake into a marsh in a matter of years. The damage to the ecology of the lake caused by siltation and water pollution is appalling. Sand mining, though advocated by some as a solution to the siltation problem, is in fact a threat to the ecology of the river systems. Apart from causing bank erosion, flooding, salt water incursion, and pollution, it is also causing drought in the valleys. A combination of these factors has adversely affected the water ecology of the state, exposing fish to diseases, mass mortality, and loss of habitat.

The indigenous fishery for mahaseer, the major game fish of the state, is reportedly dying out. Damming and channelisation activities are virtually killing the rivers of Kerala. The Periyar has been dammed at fourteen sites impacting its water flow such that in several places the river has become virtually invisible. The fate of the Bharathapuzha is even worse, with a larger number of dams cutting its tributaries. The river is dry in the lower reaches in summer, making it an easy prey to sand mining.

CAUSES OF DEGRADATION

The rural environment has been straining to cope with anthropogenic demands on its services, both as a source and a sink, for the last few decades. As seen from some of the parameters reported in On one hand, as human productive capacity has gone up, whether due to the green revolution or rapid industrialization, so has its ability to generate waste.

On the other, there is a growing demand on nature's ability to provide life support as the population keeps growing and livelihood opportunities decline. We could look at this double squeeze on nature in the context of water resources. Water applied to the field in irrigation either seeps through to underground aquifers, or reappears as 'return flow' and finds its way back to the surface (regeneration).

Pesticides, effluents include pollutants and untreated sewage, they find way into water bodies, which in turn leads to declining biodiversity. Excessive pressure on the environment leads to drought-proneness in certain areas owing to declining water table levels and flood-proneness in others owing to silting of reservoirs and loss of forest cover. A much generalized cause of environmental degradation is the failure of the governments to formulate appropriate policies to ensure sustainable land and water use. Such policy failures include price distortions through government controlled prices, subsidies or taxes which give incorrect price signals, faulty delineation of property rights regimes and other legal structures, government projects which directly cause environmental damage, and weak public institutions.

Furthermore, state appropriation of property rights has undermined traditional (often communal) property regimes, as in the case of our forest policy, and has in several cases led to de facto open access and resource degradation. Some specific causes for environmental degradation are discussed next.

Urban Spillovers and Industrial Pollution

Rapid industrialization and unplanned urban growth have resulted in heavy discharges of toxic chemical effluents and untreated municipal wastes to various water sources including streams, rivers, and tanks causing serious damage to water quality and contamination of groundwater. As human waste is flushed away into sewers and released in waterways, solid waste is collected and dumped beyond urban boundaries.

While smoky fuels have been replaced by electricity leading to radical increases in fuel efficiency as well as declines in respiratory hazards posed, it cannot be denied that this has shifted the location of environmental stress away from the city to its fringes without any policy counter-measures to relieve this stress. The strain on the environment is manifested directly in water quality, which in turn affects the soil quality and has serious impact on crop yield as well as human health. Gastro-enteritis and diarrhoea are common occurrences in villages located in periurban fringes owing to the poor quality of drinking water.

To understand the impact of industrial pollution on the rural environment, a detailed study was carried out in a village in the Patancheru region of Andhra Pradesh. The activities of the Patancheru industrial cluster have led to pollution of all fifteen irrigation tanks in the area, with five of them being completely destroyed. Farm lands have been rendered unproductive and the water unfit for drinking. The inhabitants of the hinterland are compelled to use this contaminated water owing to the paucity of water supplied. They suffer from skin infections, joint pains, and fever. Even the fish population in the nearby Manjira river has dwindled as a result of the polluted streams flowing into it.

Rural Anthropogenic Activity

The agricultural sector accounts for over 85 per cent of the total water consumption in the country. Subsidized or free supply of power and water has resulted in over-exploitation and inefficient use. Wastage in irrigation occurs to the extent of 45 per cent by seepage through unlined channels and about 15 per cent due to excess application. Losses in storage are estimated at 15 per cent.

Inefficient use of canal water and groundwater accompanied by neglect of drainage has resulted in waterlogging and salinity. The share of groundwater in the net irrigated area has increased from a third in 1965–6 to half at present to either supplement surface deliveries of water, or to provide irrigation water when limited or no surface water supplies are available. Groundwater authorities attempt to regulate groundwater withdrawals through licensing, but do not define any limits for withdrawals.

Moreover, inadequacy of resources with the monitoring agencies leads to weak enforcement. Over-extraction exceeds recharge in areas where groundwater is used for agricultural purposes. This has led to progressive lowering of the water table affecting the economy of water. Groundwater overdraft, leading to a long-term decline in water levels, has adverse consequences associated with it. These are land subsidence, deterioration of water quality in aquifers, and ingress of saline water in coastal aquifers. Salinity is likely to render more land unfit for biomass production, especially in the semi-arid and arid regions of Uttar Pradesh, Haryana, Punjab, Rajasthan, and Karnataka. Here, rainfall and available water are insufficient to leach salts from the soil.

Application of chemical fertilizers, which has recorded an increase of 80 per cent between 1984 and 1996, has added to the salinity problem. This has led to increased pH level of the water and poor drainage. Excess salts can also lead to oxygen depletion in due course of time.

Unbalanced and excessive use of chemical fertilizers and pesticides in intensive agricultural practices has led to the degradation of soils and water. In fact, the use of pesticides has risen by 240 per cent between 1971 and 1995. Fertilizers and pesticides cause acidification and eutrophication of water. Leaching of chemicals (particularly nitrates) from fertilizers results in eutrophication precipitating excessive algal growth, which kills fish. While eutrophication is one extreme, several rural regions have a net negative balance of nutrients and suffer from a gradual depletion in the level of organic matter, a trend that is likely to continue.

Application of sodium and potassium based fertilizers leads to deficiencies of macro and micro intensity, reduced biomass production and increased suspended sediments, which, in turn, adversely impact stream flow. Maintaining the nutrient balance and preventing nutrient deficiencies is a major concern given that the increasing demand for food production will have to be met through higher intensity of cropping.

Acidification, usually by oxidation, also affects the aquatic biota, pH and water quality. It also increases surface run-off. Soil erosion has led to loss of topsoil and terrain deformation. Moreover, siltation, an off-site effect of erosion, is reducing the reservoir storage capacity by 1–2 per cent annually. Exposed soil surfaces are susceptible to erosion during rain storms. Loss of soil particles and organic matter not only leads to oxygen depletion, but affects turbidity by suspended sediments on water, and influences stream channel morphology.

In addition, credit constraints have often led to an underinvestment in suitable agricultural technology and land conservation, which in the long term has resulted in falling yields from declining land fertility. Subsistence pressures lead to an outward migration to marginal areas when no land is available.

Cropping may expand onto lands which are fragile and incapable of sustaining production even with conservation efforts. This may be the case on steep slopes with shallow soil, as in the hilly regions of the north east, which would explain the high extent of degradation in this region. The impressive growth rates achieved in agricultural production in Punjab due to the green revolution are well known.

However, recent reports reveal disturbing trends about the future of the green revolution. Productivity of rice and wheat which occupy almost 70 per cent of the state's gross cropped area has reached a plateau.Problems like declining water tables, waterlogging, and salinity have reached serious proportions. The water table in central Punjab is depleting at the rate of 0.23 cm per annum.

In some parts, the water table has risen by four to five meters between 1984 and 1994 leading to problems of salinity and water logging. Widespread deficiency of micronutrients has appeared in the soil and there is an increase in weed infestation, pest, and disease outbreak. Burning of combined harvested rice straw is resulting in serious environmental pollution. Household activities such as unhygienic sanitation practices lead to the contamination of waterbodies and the spread of disease. Similarly, the use of biomass fuel for cooking and heating can give rise to indoor air pollution that threatens health, especially that of women and young children, who spend disproportionately more time indoors than men. Over time, this leads to deforestation accompanied by a loss of vegetative cover. The fertile topsoil which is held together by the vegetation now starts eroding and lack of trees also result in a gradual lowering of the water table. Given the rate of population growth, the capacity of agriculture to absorb increased working population is limited. Thus, unless there is a substantial increase in off-farm jobs and opportunities, the pressure on natural resources as a means of sustenance will keep increasing. Over fishing and overgrazing beyond the natural regeneration rate of water bodies and grasslands are all manifestations of the extreme dependence of rural population on natural resources for survival. Moreover, with the outward migration of agriculture, there is pressure on grazing lands, leading to a competition between crops and livestock.

Vanishing Common Property Regimes

The breakdown of common property regimes due to policy failures and social and political conflicts, in combination with rapid population growth, has increased the pressure, particularly on forest resources and grazing lands. This has resulted in expansion of cropping areas and increased stocking of livestock, mostly in highly fragile or marginal lands. As 'more users' increase the ambiguity in the custody of the common property, it is overexploited and in the long run becomes completely degraded and unfit for use. With the 1980 Forest Conservation Act, tribal populations living in the forests for centuries were suddenly devoid of all use rights.

With vast tracts of forest areas being declared national parks and sanctuaries, the tribals have no option but to break the law to earn basic subsistence. Unfortunately, government response to such encroachment has been strict policing rather than leading to deterioration of forest land rather than involving the tribals in forest management. Joint Forest Management, the process of involving tribals in maintaining forests, is still limited to small pockets of the country and there is a dire need to expand such programmes.

Government Failure in Managing the Ghasnies of Himachal Pradesh

The Sirmour district in Himachal Pradesh lies in the outer western Himalayan Range. Mixed-farming is the main occupation in the mid-hills and people combine private resources, state-owned, and commonly-owned resources in their day-to-day functioning. Until 1974, grazing lands or ghasnies, part of the forest surrounding villages, river beds, irrigation channels (collectively called shamlats) and common paths belonged to the Panchayats and were used for the common purpose of the village.

On account of grass resource seasonality, customary rule imposed on graziers restricted access to grasslands for a period of time. The rule in practice was to prohibit access to the grasslands during the rainy season until it could be collectively harvested in September. Regulatory in nature, these informal restrictions were monitored and enforced by mutual agreement among the user group. The size of the herd allowed to graze the commons in summer was not a binding criterion. However, households were entitled to graze as many animals in summer as they were able to feed in winter period (when extra source of fodder is not available).

This rule aimed at preventing commercial rearing which could take place during the grazing season. Working within the limits of its labour force, a single household was entitled to remove as much as it could from the common pool resource as long as these benefits were not cash or market-oriented. In this way, there were 'social boundaries' on households' usage of the ghasnies. Due to increased concern over the environment in the 1970s, the State Government enacted legislation through which common land ownership by the Panchayat was transferred to the state unless the land was 'subjected to partition between individual co-sharers before the Act commenced'.

This led to powerful farmers securing individual access to and benefits from the commons before nationalization. This change in access to the village commons had several negative outcomes. Large tracts of the newly private pastures were increasingly converted into cultivated lands. Agripastoralists excluded from 'wild privatisation' were forced to graze their cattle in the forest with intensified green lopping on selected tree species such as Ban and Moru, which have now almost disappeared from the region. Forest lands, which are de jure public lands, are now used as open to all.

The remaining parts of the shamlat are vested with the state which has bifurcated its use into grazing lands and allotable pool land for cultivation to land-less labourers. Productivity of these lands are abysmally low. Finally, encroachments and further political regularization have been important factors responsible for both the shrinkage of grazing lands and consequent intensive use of forest for lopping and grazing.

This example illustrates how inappropriate policy can have adverse ecological effects. The de facto partition of common grazing lands has resulted in a reduced commons area, which led to overgrazing and gradual destruction of forest lands. Moreover, control by the state has resulted in numerous possibilities for the individual to escape both legal as well as customary systems of sanctions which makes the rules-in-use unviable. Under these circumstances, collective action can emerge as an efficient alternative to 'privatization' and 'nationalization'.

Poverty and Environment

Natural resources are the most significant source of rural livelihoods in the country. Not surprisingly, some of the poorest regions in the country are also places where agriculture is least developed and land and water resources are poorly managed. Conventional thinking on environment blamed the poor for over-exploitation of natural resources, as poverty and environment were considered linked in a 'downward spiral' in which poor people, forced to overuse environmental resources for their daily survival, are further impoverished by the degradation of these resources. Population growth and economic change (which often bypasses the poor, or reduces their access to natural resources) are also seen to contribute to this process. It was, therefore, believed that poverty needs to be eradicated in developing countries before they can turn their attention to environmental protection. However, the perception of the 'vicious circle' as characterising the environmental degradation and poverty in developing countries is somewhat simplistic and misleading. For example, if one looks at history, there was not much environmental degradation even when poverty levels were much higher.

Now that poverty levels are declining significantly, it does not seem plausible to attribute environmental degradation to poverty. Evidently, other factors such as property rights, access to resources, use of traditional wisdom, and misappropriation of environmental resources in the name of development play a more important role. As the poor are dependent on nature for their livelihood, they are very vulnerable to natural calamities, environmental degradation and ecological disasters (which are often man-made, like the Bhopal Gas Tragedy, and pollution caused in the river Yamuna by industries and upstream rich farmers).

There are cases of destructive development which aggravate poverty and contribute to environmental degradation at the same time. Thus, there is enough empirical evidence now to establish that environmental conservation must go hand in hand with economic development because any economic development which destroys the environment will create more poverty, unemployment, and disease. However, it is disturbing that neither concern for establishing equity nor the desire to preserve the environment underlies the recent interest in sustainable development among the most developed countries.

Rather, the primary concern continues to lie in sustaining their consumption levels and in maintaining the conditions necessary for economic growth. In contrast, the environmental priorities of developing countries are underpinned by grim and undeniable human realities. In many rural areas in India, environmental issues are issues of life and death. The protests of the people in these areas are responses to social injustice and signal the rising political and ethical consciousness in rural India. In fact, if poverty is indeed the major cause of environmental degradation, the poor actually have a good reason to support conservation movements.

Unplanned Development Projects

The environmental impact of unplanned development projects which includes construction of dams, taking over of huge tracts of land for mining, or converting natural forests to forest farms for wood pulp generation, vary in intensity but have some common and inescapable consequences. When implemented without proper cost-benefit analyses these projects result in millions of oustees, violent disturbance of pristine areas, varying degrees of submergence of land including forest land, adverse impact on flora and fauna leading to a reduction in bio-diversity, and severe impact on the fish population in rivers. Stilling of flowing waters leads to temperature stratification, variations in nutrient content and dissolved oxygen, rendering the water inhospitable to aquatic life.

Building dams, in particular, results in drastic changes in the downstream river regime, including reduced flows affecting aquatic life and riparian communities, reduced capacity for self-regeneration, increased pollution levels, reduced recharge of groundwater aquifers, and adverse impact on estuarine conditions. Acquiring land for development projects is not a new phenomenon and resettlement is usually limited to giving cash in exchange. The displaced are then left to go ahead with their own rehabilitation plans.

These people live at the mercy of nature and disease. Their appalling living conditions, housing, health, and nutrition are an insult to the notion of equity. There need to be proper models of analysis that focus on mitigating risks and formulating alternative strategies. There should be greater participation of local people in the decision making process. Only when people have the power to influence the decisions of the management from the very initial stage will development activities stand a long-term chance of success. The draft Scheduled Tribes (Recognition of Forest Rights) Bill 2005 would be a step in the right direction if it were to be passed as an Act.

MANAGING THE ENVIRONMENT

Policy Changes

Environmental concern in India is inextricably linked with peoples' lives and well-being. Several parameters currently exceed threshold limits of tolerance to the extent that some environmental problems pose real and immediate threats to health and livelihood. Given that the most serious environmental issues in rural areas emanate from improper agricultural practices, it is essential to integrate agricultural and environmental policies.

Specific policies targetting environmental conservation milestones should be built into sectoral and macroeconomic policy targets. Unless the environmental checks and safeguards are institutionalized with the provision of other rural infrastructure such as roads and communication, the socio- economic condition of rural people cannot improve in a sustainable way. While poor environment undermines development, inadequate development results in a dearth of financial resources which can be devoted to environmental protection.

The policy approach to agriculture, particularly in the 1990s, has been to secure increased production through subsidies in power, water, and fertilizer, and by increasing the minimum support price. The creation of new capital assets in irrigation, power, and rural infrastructure has not received priority. The fertilizer subsidy is, in fact, meant to cover the high cost of the fertilizer industry. Excessive use of subsidized fertilizer has created an imbalance between Nitrogen, Phosphorus, and Potassium in the soil while excessive use of water–given the 'free' availability of power–has resulted in water logging in many areas. Under-pricing of power and irrigation, rather than bringing in improvements in rural income distribution, has led to inefficient utilization, with negative environmental externalities. The subsidies have grown over the years and are now financially unsustainable.

With deteriorating state finances, subsidies have 'crowded-out' public agricultural investment in roads, irrigation, and expenditure on technological upgrading. The equity, efficiency, and sustainability of subsidized power and fertilizers are, therefore, questionable.

Awareness and Attitudes

Although Environmental Impact Assessments (EIAs) is now a prescribed requirement for all projects, in reality they are often inadequate and unreliably conducted studies, making their cost-benefit analysis a suspect. Even the statutory requirement of a public hearing in relation to the environmental clearance of a project is yet to become a well-established procedure. Awareness and advocacy among the local people is important as they are direct stakeholders in the implementation process of these projects, and community capacity-building is necessary to ensure that these people can seek legal recourse, if required.

The answer, however, does not lie in large, centralized, 'topdown', technology-driven projects: local, decentralized, community-based, people-centred alternatives are available. Problems of scarcity of water, depleted aquifers, declining groundwater tables, and drought proneness have been successfully tackled by water harvesting endeavours in Ralegan Siddhi village in Maharashtra, Sukhomajri in Haryana, and Alwar in Rajasthan.

These are not 'small' instances to be dismissed as one-off phenomenon but examples of significant and sustained success achieved in terms of increased water availability and rise in groundwater table. If these examples could be replicated across India, they could act as key drivers in the process of environmental restoration. Larger projects, also, cannot be ignored, as there are circumstances where benefits from micro-level endeavours may not be enough. The possibility of integrating a large project with smaller surface and underground storages within the sub-basin or watershed must be considered. A negotiated participatory approach should be adopted in ensuring success of any public intervention, involving communities and land users in creating and maintaining environmental infrastructure. This will leverage local experience and knowledge and ensure maximum involvement. Decentralization in one form or another has been the primary vehicle for bringing democracy at the local level to facilitate local control.

This implies either redistribution of authority from the central government to field delegation (de-concentration), the transfer of specific decision-making and management authority to semi-independent units (delegation), the transfer of authority to nongovernmental sectors (privatization), or strengthening and transfer of power to local governments (devolution). However, there are practical difficulties associated with each of these measures. To manage forest resources for example, community biodiversity registers must be developed before issues of ownership of resources and equitable sharing of benefits are sorted out. Moreover, clear and unambiguous guidelines (to pre-empt any possibility of rent seeking) for managing bio-prospecting agreements need to be drawn up keeping in mind the interests of the locals who derive their livelihood from the forest.

Local communities can be involved in forest conservation by employing them as wildlife guards and forest watchers and by using them as trainers to build anti-poaching awareness among their localities. Indigenous skills can be utilized by employing locals for collecting snake venom and capturing and training wild elephants who devastate standing crops.

Water Harvesting in Alwar: Revival of the Tradition of Johad

A Johad is a dug-out pond, created at a place chosen with native wisdom, informed by remembered patterns of water flow during the rains to harness the rainwater run-off with high embankments on three sides. The height of the embankment depends on the volume of run-off from the catchments. The water storage area varies from 2 hectares to a maximum of 100 hectares.

The water collected in a Johad during monsoon penetrates into the sub- soil and recharges the groundwater, improving soil moisture in vast areas mostly downstream. Apart from arresting and storing rainwater, it stops soil erosion, mitigates flood, and ensures water availability in wells for several successive drought years. The groundwater can be drawn from traditional open wells, built and maintained by the villagers themselves.

The water from the Johad is also directly used for irrigation, watering of domestic animals and other household purposes. During the dry season, when the water gradually recedes in the Johad, the land inside the Johad becomes available for cultivation. This land, by receiving good silt and moisture, allows crops to grow without irrigation.

Johad is built using simple technology and local materials. In the Alwar district of Rajasthan it took three years to build the first Johad. In the fourth year, Tarun Bharat Sangh, a nongovernmental organization (NGO) actively helping villagers, had built fifty Johads. As on date, 9000 such structures exist catering to water needs of more than 1000 villages.

This area which was classified as 'dark zone' in 1995 was reclassified as 'white zone' in 2005. As water availability improved, agriculture became productive and cattle rearing started in earnest, resulting in increased production of milk. Studies have shown that an investment of `100 per capita on Johad raises village domestic production by `400 per capita per annum.

Non-farm Employment

In addition to policy changes to improve agricultural processes, expansion of the non-farm sector is essential for development. If rural non-farm and urban informal sectors grow at an adequately high rate to absorb the surplus labour and surplus food then the scenario is a positive one. If however, they grow at a lower rate, the terms of trade can turn against agriculture leading to depression of agricultural incomes.

Rural enterprises sector can, in this context, become both an engine of growth as well as a major contributor to the reduction of rural poverty. The potential for labour absorption is high in agro-based industries, small- and medium-scale industries, and the rural services sector. Most of these activities are highly labour intensive and provide employment opportunities for the semiand unskilled rural labour and semi-skilled urban labour force.

When non-farm activities provide stable sources of income to small farmers, they can afford to opt for high-value crops with higher associated risks. Income diversification of farm households and the insurance role of non-farm activities would strengthen the viability of a small farm economy. Thus the income of farmers can be magnified by multiple linkages with the non-farm sector. These are production linkages, both backward, via the demand of farmers for inputs such as ploughs, engines and tools, and forward, via the processing needs of agricultural goods through spinning, milling, canning, and so on.

Table. Growth of Employment in Rural Enterprises

State/Union Territories	Annual growth rate in non-farm rural employment	Annual growth rate of rural workforce
Andhra Pradesh	5.37	3.05
Arunachal Pradesh	3.65	3.07
Assam	6.62	2.08
Bihar	4.50	1.79
Chhattisgarh	3.24	3.82
Goa	1.75	2.99
Gujarat	3.11	1.27
Haryana	9.68	8.80
Himachal Pradesh	2.73	2.54
Jammu and Kashmir	7.64	7.65
Jharkhand	3.44	0.66
Karnataka	4.78	2.69
Kerala	7.93	4.21
Madhya Pradesh	1.74	1.69
Maharashtra	4.95	3.29
Manipur	4.46	3.24
Meghalaya	6.48	5.05
Mizoram	8.40	4.96
Nagaland	6.05	1.95
Orissa	3.02	2.54
Punjab	7.34	5.19
Rajasthan	4.15	3.44
Sikkim	8.39	6.41
Tamil Nadu	9.96	5.43

Tripura	9.85	5.84
Uttar Pradesh	7.07	4.98
Uttaranchal	7.72	7.06
West Bengal	4.77	1.70
Andaman and Nicobar Islands	-6.16	-3.90
Chandigarh	15.57	12.11
Dadra and Nagar Haveli	8.65	7.56
Daman and Diu	13.64	15.32
Delhi	-0.91	-2.26
Lakshadweep	1.80	3.53
Pondicherry	3.37	3.83
India	5.53	3.33

Consumption linkages are established with higher agricultural income feeding into an increased demand for goods and services produced in nearby villages and towns. Improving rural connectivity also has an impact on creating employment opportunities. It is fairly common to see tea stalls, bicycle, and tractor repaïr shops and other establishments clustered around bus stands. Such options can ensure to an extent that rural populace does not need to exploit natural resources for subsistence, contributing to a healthier rural environment.

The annual growth of non-farm rural employment for 2005 exceeded the average annual growth rate of workforce at 2 per cent per annum. The average annual growth in employment for the years 1998–2005 was 2.5 per cent compared to the rate of 1.7 per cent between 1990 and 1998.

While this is heartening, it is important to continue with the thrust on rural employment, through the wide spectrum of employment programmes devised by the central government. In addition, there is a need to increase the value added component of non-farm rural enterprises so as to improve their profits and make them sustainable.

Contract farming facilitates the integration of small farms with agro-processing companies, which extend technical expertise and financial support to the farmers to grow high value crops. Historically, contract farming has been a standard practice in the Indian sugar industry and could be extended to other product groups to trigger the development of agro-processing industries and generate employment for surplus rural workforce. Further, commercialization of agriculture could catalyse growth rates for the sector in a way that agricultural development could diversify into dairying, animal husbandry, fisheries, floriculture, horticulture and other areas. This would spur the growth of agro-processing industries in rural areas to meet domestic as well as export demand.

The tremendous potential of the khadi and village industries to generate employment in rural areas to ease the increasing pressure on the land could be exploited by developing forward looking plans and overhauling the marketing and distribution networks. The lacunae are in the formulation and implementation of the market development plans.

The government or NGOs could play a more proactive role rather than just engage in procurement and selling through Khadi and Gram Udyog network. It is essential to make khadi and village industries sustainable and economically viable. In this context, the focus of the policy should shift from providing subsidies to market development.

Decentralization

The notion implicit in the decentralization process is that effective management of local resources requires effective participation of individual users at the community level. People's participation in the decision-making process is an important component of decentralization. The objective of the devolution of power is to involve the target groups of state developmental efforts in a very real sense in the formulation and management of developmental projects and programmes.

The 'local people' is a heterogeneous group defining the primary consumer set for local services. This group is the key stakeholder in the success or failure of grassroot programmes as well as civic and administrative systems in place at the local level. Panchayats, as the form of local government, have existed as a traditional village institution for centuries. In 1992, the 73rd Amendment to the Indian constitution identified areas where these grassroots level selfgovernments could be granted administrative power in 27 areas including agriculture, irrigation, and management of drinking water.

This policy shift towards decentralization promises important social change in rural India, providing as it does a three-tier system of Panchayats: at the village level, the district level, and an intermediate level between the two, called the Block Panchayat. The devolution of power to the Panchayat in the management of natural resources, such as water, is a shift from the centralized and state-driven natural resource management regimes of the past towards decentralized and mainly community based management.

Instead of controlling resources from outside, decentralization could enhance participation in the form of devolution of decision-making power to formal institutions, such as, local governments that possess meaningful autonomy and the capacity to mobilize and spend resources. In particular, if there is a traditional institution of decision-making, then it can be easier and more expedient to use this as a participatory channel rather than creating new committees or institutions.

The whole process, in an ideal situation, enables the local people to debate, negotiate, and have control over the resource. If this really happens, then with time, local governance can become an educative process where the participants learn about democratic norms and practices. Because of the dominant role of natural resources in local livelihoods, it is true that people need to have an effective voice in decisions over the natural resources they depend on.

The proponents of decentralization argue that the establishment of local (formal) institutions has the capability to improve people's management and use of common property resources, thereby improving the resource base on which poor people are often disproportionately dependent. It is hoped that through these institutions, participation can better target benefits to the poor through the identification of key stakeholders who are most affected, and can imply an on-going information exchange and discussion through consciousness-raising by shared understanding of problems and a vision for the future that leads to commitment and ownership by the community.

The governance structure, it is hoped, will change as a result of decentralization from centralized to localized, with the 'people' at the centre. Ideally, the higher authorities will not manage natural resources, but through a participatory process, the local people will manage them, thus resulting in a change in the pattern from a 'command and control', to a 'responsive and accountable' operative system.

The new people-centred bottom-up paradigm in development thinking has created the overly optimistic view that decentralization will produce just and equitable outcomes for all, and that engaging the people will also act as a check on state power, thus helping to democratize local governance. The new paradigm stresses the involvement of local people in contrast to the top-down paradigm, and tends to dominate management of natural resources at the local level. Chambers argues that the emergent paradigm for humans living on and with the earth brings together decentralization, democracy, and diversity. What is local, and what is different, is valued. In this paradigm, the trends towards centralization, authoritarianism, and homogenization are reversed. Reductionism, linear thinking, and standard solutions give way to an inclusive holism, open systems thinking, and diverse options and actions.

The importance of traditional ways of combating with problems could be important too: here, informal institutions could be involved. For instance, whenever villagers in Karnataka's Bijapur district sense a drought is imminent, they prepare for war with nature. Harbingers travel from place to place and try to bring rain through magic. Rainmaking may not work but the participants at least endeavour to do something in a situation that would otherwise leave them dispirited and helpless. The accompanying sense of accomplishment enables the locals to cope with drought; it is preferable to sitting idle and wringing one's hands.

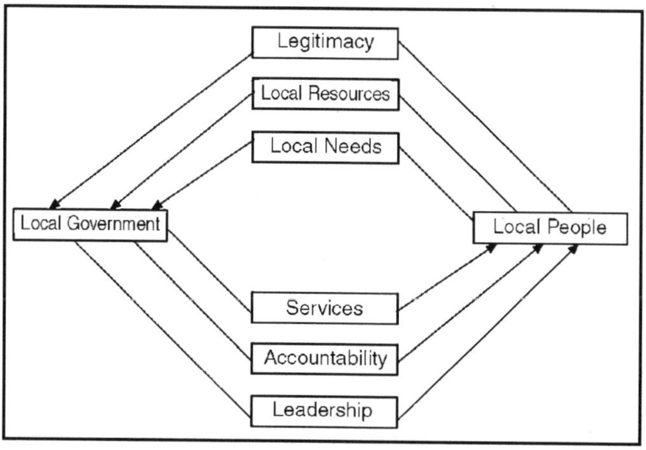

Fig. 5.1: The ideal situation: Towards a Good Governance for Natural Resource Management

Besides the above changes in policy orientation, there is also a need to revamp the existing institutional set-up responsible for rural environment. The current policy environment suffers from lack of coordination and communication between different departments and agencies, the activities of which impinge on natural resources.

A multi-level stakeholder approach for the planning process is essential to obtain socially balanced results in which both the economic and ecological objectives are given due weight. At the district level, a coordination committee headed by the district collector, involving the forest department and other relevant agencies including the District Rural Development Agency could be created to oversee the impact of different projects on natural resources.

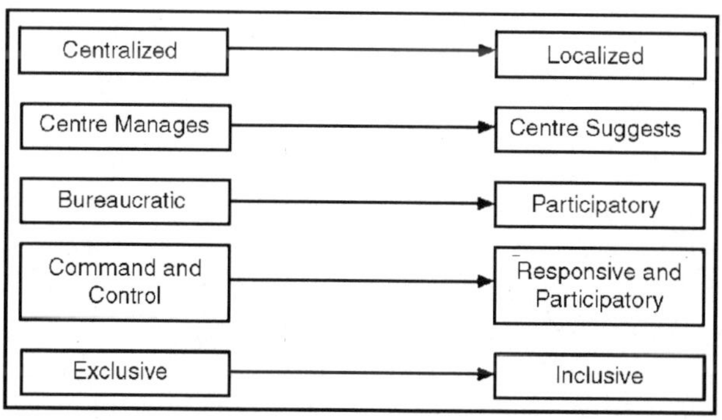

Fig. 5.2: Possible changes in Natural Resource Governance because of Decentralization

Similarly, to ensure that all development activities and actions relating to the utilization, production, or management of natural resources are evaluated before implementation, it may be necessary to create an environmental cell in all the departments with expertise in the management of air, water, land, and biodiversity in the face of developmental compulsions.

Isolated planning and implementation of surface water and groundwater development programmes discourage unitary analysis of water and its use as a single resource, resulting in duplication and ambiguity of functions. Central and state groundwater boards have their own separate observation wells for monitoring water tables, but do not share the data among themselves, and doubt each other's databases. At the same time while the Central Pollution Control Board and the state pollution control boards focus on pollution from industrial and domestic sources, agricultural pollution is overlooked. Institutional transformation towards granting operational autonomy in the irrigation sector is called for in conjunction with the establishment of a multidisciplinary environmental unit to encourage the conjunctive use of surface water and groundwater, for irrigation, which would ensure better drainage in canal command area, thereby checking water logging and salinity. Similarly, the utilization of treated wastewater from domestic and industrial sources for irrigation could be promoted. Water management should also be closely synthesized with land management to tackle degeneration of ecosystems regulated by groundwater and surface water.

This should be a crucial aspect of institutional reform along with three other major issues: planning and operation of large surface systems, watershed management, and regulation of groundwater. Indeed, groundwater is one of the country's most valuable natural resources. Despite the urgency to rapidly develop new water supplies, little attention has been paid to or investment made in the preservation of the long term sustainability of groundwater.

The National Water Policy had recommended the creation of multidisciplinary units in different states for water management and proposed participation of beneficiaries in water management and water price rationalization among others.

However, water management is yet to take off in many states. A major drawback of environmental institutions is that they are slow to incorporate appropriate policies amid a dynamic situation of technological and social change. Environmental policy-making and administrative and implementation mechanisms are subject to the inertia of the historical *status quo* of special interests.

Improving Managerial Efficiency

While policy and institutions can enable environmental restoration, solutions are often found in technological breakthroughs or suitable managerial interventions. In the following sections, we highlight some of these solutions towards monitoring environmental degradation, managing natural resources, enhancing agricultural growth, and specifically rejuvenating the rural environment. It must be noted that while these have been organized as separate sections for the sake of readability, the solutions are interrelated and cannot be implemented in isolation from each other.

MONITORING

There is a rising concern to examine and monitor the changing health status of soil, land, and water in the rural hinterland. To start with, it is necessary to take stock of the current environmental situation by studying data on the properties of the land, different types of land-use options and their effect on the resilience of ecosystems. Technology transfer options have to be assessed in the context of training farmers, especially the small and marginal.

Projected increase in cropping intensity and consequent nutrient depletion demands the establishment of a monitoring mechanism for soil nutrients. Land capability classification to ensure efficient resource allocation entails completion of land and soil surveys and the development of an inventory of land resources on the basis of land-use classification.

The main objective of collecting land use statistics in India has been to facilitate agricultural accounting. However, the rise in human and livestock population and development of modern agriculture have led to degradation of land, which, if unchecked, may result in irreversible damage. It is important, therefore, that a systematic database be built to assess changes in land use and the type and extent of environmental damage to land resource in each use.

Land use database has to be significantly expanded from revenue and agricultural production assessment to assess and monitor the changes in the ecology and the environment. Data on barren and uncultivated land, already collected by the village Patwari, need to be aggregated as non-agricultural uses have increased substantially and the data would enable environment monitoring and decentralized planning at the district and block level. Soil quality data relating to topsoil erosion, salinity, drainage and waterlogging can be routinely monitored and remediation action initiated wherever needed.

NATURAL RESOURCE MANAGEMENT

Combating pressure on forests would require demand-side management with respect to wood and fodder and supplyside management with respect to enhancing wood yield in forests and increasing the supply of fodder. 'Open forests' representing 26 Mha are already degraded. According to Pachauri and Mehrotra, 15 Mha of this can be managed under the joint forest management (JFM) programmes and the remaining used for commercial timber production.

In fact, nearly 10 Mha is already under the JFM leaving only a remaining 5 Mha to be covered. Of the degraded land earmarked for commercial timber production, 4.2 Mha of scrub (less than 10 per cent cover) can be brought under plantations and 4 Mha currently under cultivable waste and permanent fallow could be used for raising trees. This would imply an increase to about Mha under tree cover by 2020.Plantation forestry requires the participation of communities, NGOs, and the private and corporate sectors.

There is a need to increase the area under agro-forestry and farm forestry on agricultural lands by planting trees along farm boundaries and in homesteads in a manner that does not affect crop production adversely. Strategies for promoting agro-forestry should focus on removing legal barriers, bringing about market reforms, fostering farmer–industry linkages, and enhancing the overall profitability of farm forestry.

In addition to managing forest resources, there is a need to manage the requirement of fodder for farm animals and land devoted to pastures. Limiting livestock population through selective breeding and pasture and fodder management could mitigate the pressure on land and forests. Moreover, specific forage resources could be created for livestock so that they do not compete for productive agricultural land. A detailed assessment of fodder requirement at the national and regional levels could be carried out on the basis of which regional allocation for production of fodder can be undertaken.

Encouraging stall feeding to control grazing, providing incentives to farmers to include green fodder in crop rotation, and using crop residue as fodder are alternatives that may be considered. Fodder could also be grown on land under agro-forestry and JFM. Reduced demands on forests for grazing in the long term coupled with wood-demand optimization and strategic forest management will mitigate the pressure on forest resources. This, in turn, will have a positive impact on the country's biodiversity.

AGRICULTURE

Farming systems vary with agro-ecological conditions and no single technical fix will work as a magical cure for improving farm productivity. In some regions, solutions for increasing yields may involve a shortening of fallow periods and extension of cropping periods while in others where soil fertility and/or access to purchased inputs is good, solutions such as annual cropping or multi-cropping without fallow would work.

Again, farming systems based on tree crops, such as tea and coffee, are suitable for some regions only and should be encouraged accordingly. Further, the degree of market integration, choice of crops and cropping systems, use of conservation technologies, and use of purchased inputs and their effects on the farming system, are all important in determining the sustainability of particular farming systems.

REVAMPING AGRICULTURAL EXTENSION

Revival of agricultural dynamism calls for corrective steps to deal with the near collapse of the extension systems in most states and the decline in agricultural research universities. Laboratory-to-land concept should be encouraged and put to practice by providing land-users multidisciplinary technical information and viable land-use options and alternatives identified for various agro-ecological and socio-economic units. Crop combinations and rotations suitable for different agro-ecological regions (as suggested by the Indian Council of Agricultural Research) need to be advocated for better land management.

There is a need to stay abreast with evolving resource conservation technologies and practices and on analysing the conditions and principles of sustainable land use. Efficient use of marginal lands needs to be encouraged and areas of untapped potential developed to ensure optimal utilization. For agricultural diversification to be a major element in the agricultural growth strategy, action on several fronts is called for. Ideally, there should be a shift of land from cereals to non-cereals (increasing both farm incomes and employment) combined with an increase in productivity in cereals to ensure that per capita availability of cereals does not decline.

Improvement in fertilizer application efficiency, integrated with the use of bio fertilizers, to check the degradation of existing resources due to contamination with nitrates could be brought about through on-site farmer training programmes. Success in providing extension services so that breakthroughs in research can be implemented by the farmers entails implementation of extensive irrigation reforms.

BIOTECHNOLOGY

While the possibilities of application of biotechnology to improve the rural environment are limitless, we focus on two specific applications: the improvement of genetic stock of farm animals and forests. Other applications of biotechnology in food and non food-crops like tomatoes and cotton are well known.

REJUVENATING THE RURAL ENVIRONMENT

Salinity

Soil salinity has been emerging as a critical problem in both productive and non-productive rural land. While social forestry is a popular means of tackling salinity innon-productive land, salinity management plans in catchments, pastures, and farm areas include a combination of prevention, recovery, containment, adaptation, and pursuing sustainable agricultural practices so that salinity is mitigated over time. The prevention approach involves retaining native vegetation on the land as far as possible to avoid the risk of salinity. Thus, the migration of farmers outwards to marginal areas must be discouraged to retain natural vegetation which can be used for managed grazing of livestock. The recovery approach brings saline land back to full production under its former land use, although this can entail high costs based on currently available solutions. Containment strategies focus on vegetation management so as to reduce the rate of increase in water tables and avoid the spread of salinity. Adaptation refers to changing the land use pattern itself because the soil is saline. This is a good option when there are no cost effective measures for soil recovery and alternate land use such as growing feed for livestock is more profitable than growing crops in the degraded land. Deciding on the right approach to tackle salinity is a measured step combining the interests of ecological balance with socioeconomic needs of the farme. Social forestry has been the most cost effective method to tackle salinity so far. Trees are planted to reduce groundwater recharge, to intercept and use groundwater before it reaches a discharge site, and/or to increase the rate of discharge at a saline site. These approaches work best when they are designed to suit the hydro geological characteristics of the system causing the salinity.

Trees have a higher potential to reduce recharge (compared to that under annual crops and pastures) and to intercept and use groundwater. Strategies that could maximise the reduction in recharge or use of groundwater for a given area of farm forestry include planting on high recharge zones; planting to use groundwater, for example, having tree belts across hill slopes, break- of-slope plantations, and plantations bordering saline water discharge zones.

Trees near annual crops and pastures may be able to reduce the percolation and recharge below the nearby annual plants if they draw water from a zone wider than the tree canopies. This is usually done by planting rows of trees across paddocks, leaving alleys for conventional farming. Also, trees planted on sites where large soil water stores have developed below annual crops and pastures may be able to access this water with their deep roots and prevent it from becoming recharge; this is referred to as phase farming. Trees can also use surface water flowing onto their site if it infiltrates and is stored below ground within reach of their roots.

Rainwater Harvesting

There are many arid zones and drought-prone areas in the country in Rajasthan, Gujarat, Maharashtra, Karnataka, Andhra Pradesh, and Tamil Nadu. Droughts are a recurring feature in these areas, causing untold misery to human beings and livestock, often resulting in large-scale migration. There needs to be a series of area-specific answers by way of local conservation and augmentation to the maximum extent possible to drought proof these areas.

Rainwater harvesting as a method of utilizing rain water for domestic and agricultural use is already widely used throughout India and the world. It is a method which has been used since ancient times and is increasingly being accepted as a practical method of providing potable water. It has wide application also in peri-urban areas where the reliability and quality of piped water is increasingly in question. With most parts of the country subject to a large variation in rainfall leading to spatial and seasonal variation in water supply, rain water harvesting is an ideal tool to take care of local water needs as well as restore the ecological balance of regions through groundwater recharge. Success stories of rain water harvesting abound in villages in different parts of India, a compendium of which can be found at a web site for rain water harvesting maintained by the Centre of Science and Environment.

Common Property Management

Common properties are best managed by social institutional arrangements that enforce a structure of management. Fundamental to understanding the complexities involved with management of a common property resource is the way the government (or the people themselves) form(s) institutions around the resource. The institutions which control decisionmaking regarding natural resources play an important role in providing access for all.

Small group size, well defined boundaries on resources and user groups, and ease of monitoring and enforcement are essential conditions for the success of these institutions in sustaining common property resources and gaining the compliance of generation after generation of appropriators to the rules in use. While informal CPR management arrangements, being socially embedded, have the potential to solve problems of access, they need to work within the framework of formal organizations.

Formal rules could determine the organizational structure and function of a particular informal institution managing a natural resource, and rules and sanctions, while informal arrangements can ensure involvement of locals and sustainability of conservation efforts through cooperative agreements. These organizations are in a better position to analyse the processes involved in conscious inclusion (or exclusion) of certain people in the use of a natural resource.

Of the three solutions that exist to resolve this problem of tragedy of the commons, namely centralized management of the resource, privatization of the resource, and local control of the resource, local control is preferred because it is more inclusive than the other options.

Making local 'user groups' control the common property resource by giving opportunity to people who have a common material interest to participate in management of the resource has been found to be both ecologically and economically efficient.

'Water user organizations', for example, are becoming increasingly popular in India, and are more efficient than the government machinery for managing water resources. However, the application of informal CPR management has, in many ways, been unduly restrictive. It has been focusing on micro-enterprises in management of forest patches at the village level, rather than looking at higher-level forest management and economic processes.

The emphasis has been on the bare bones of participatory development— new rules, one-off participatory planning events, village committees–rather than broader and longer term processes such as the management of trade-offs among diverse objectives, and the need for scaled-up participation. Production of trees, tree products, and fodder has been emphasized upon while broader ecological processes and debates about alternative uses and competing rights and responsibilities have been overlooked.

Information Systems and Accountability

Information Technology and techniques that are freely available, can minimise errors and provide accurate and reliable periodically updated data. Two related issues are the cost and the system of storing information. Currently, land use data is generated, updated, and maintained by the Patwari at the village level. The information is generated as a by-product of administration and hence, has a low marginal cost. Storage of information is not an issue. However, dissemination of the data should be made as simple as possible so that all departments have access to all the information, which can be viewed holistically before undertaking projects that could impact the environment.

Remediation of Wasteland

Remediation refers to conversion of wasteland which has sustained environmental degradation into land suitable for habitation or cultivation. Approximately 20 per cent of the geographical area of India is now under the wasteland category. Growing demand for fuel, fodder, wood, and food has extensively depleted or eliminated protective plant cover and exposed surface soils to processes of degradation, resulting in partial to complete loss of soil and productivity. As a result the production of vegetation for food and other uses has extended to areas under great ecological stress and with less favourable environment. Several approaches to land remediation have been successfully implemented. For lands that have not sustained too much damage, measures such as building suitable rain water harvesting structures, planting appropriate vegetation and trenching of soil help in improving land productivity. The age old practice of leaving lands fallow for some seasons is also a means of improving productivity of the soil. Jatropha, a valuable crop, owing to its capability in producing bio diesel, is an option which can be chosen to alleviate soil degradation, desertification, and deforestation. The concept of substituting bio-diesel produced from plantations on eroded soils for conventional diesel fuel is gaining acceptance in India.

In recent months, the centre and several state governments have expressed their support for bringing marginal lands, which cannot be used for food production, under cultivation for this purpose.

Management of the Sundarbans

The Sundarbans, home to the endangered Royal Bengal Tiger, cover about 1 million hectares of land and water across Bangladesh and West Bengal, comprising one of the three most extensive mangrove forests in the world. They are the nursery for many coastal and oceanic species of fish. Bio-diversity in the region includes 36 mangrove species, turtles, terrapins, river dolphins, ies of birds, many of which are migratory. The Sundarbans are also among the most backward regions in West Bengal. This region is home to about 4 million people, of whom 56 per cent are landless. Literacy rate here is below 35 per cent, and most communities do not have access to electricity or safe drinking water, and agricultural productivity is lower than the state average. Inadequate infrastructure, poor communication facilities, lack of access to clean drinking water and health and education services, and a fragile and limited natural resource base have contributed to a low level of development and high poverty incidence in the region. A diminishing natural resource base is threatening the ecological integrity of the Sundarbans and the livelihood of the inhabitants. 90 per cent of the locals, who depend on agriculture, draw heavily on forest resources to meet their subsistence needs. Thousands of women and children collect wild tiger prawn spawn for sale to shrimp farms, leading to a decline in tiger prawn spawn and concurrent decrease in other fish species. The frequent breaches in the embankments along the riverbanks built to prevent inundation by tidal waters allow saline waters to inundate agricultural lands, destroy crops, and render the soil infertile.

The vulnerability of the region to periodic cyclones forces the local communities to turn to the forest to meet their food and shelter needs. This has resulted in over extraction of wood and non-wood products such as palm leaves, honey, and fuel wood. Low material and technological inputs are constraining agriculture, crab farming, bee keeping, and honey production.

Several initiatives were taken in the past few decades to enhance conservation and development in the area. In 1973, 2585 square kilometres of mangrove forest was declared as a Project Tiger Area. In 1976, three wildlife sanctuaries were created, and in 1984, most of the core of the Project Tiger area was converted into the Sundarbans National Park. The Sundarbans Biosphere Reserve (SBR) was established in 1989 over a larger area.

These initiatives are backed by an enabling policy and regulatory framework, which provides for community involvement in mangrove forest protection and management, and for regulating fisheries. However, effective enforcement of rules and regulations suffers from lack of inter-agency coordination and insufficient enforcement capacity. The extreme poverty in the area led the state government to form the Sundarbans Development Board (SDB) in 1973.

The Sundarbans Affairs Department (SAD) was created in 1994 to give further impetus to socio-economic development in the area, and SDB is now under this department. During 1981–1990, SDB implemented a US$ 9 million integrated rural development project funded jointly by the International Fund for Agriculture and Development and the state government. In 2000, SDB's US$ 4 million development budget focused on civil works and infrastructure development, afforestation, agriculture, fisheries, and small-enterprise development.

The Forest Department (FD) implements central and state-sponsored schemes focusing on mangrove forest rehabilitation, fuel wood and fodder development, and community development. Other state departments including irrigation, public works, public health engineering, and several nongovernmental organizations (NGOs) focus on socio-economic development, mangrove forest restoration, and wildlife conservation.

The United Nations Development Programmes and World Wide Fund for Nature have proposed initiatives to focus on reproductive health and population issues related to Sundarbans conservation; fostering stronger cross- border cooperation, and collaboration between conservation managers in India and Bangladesh; and tiger conservation.

Yet another alternative is the use of mycorrhizal fungi for soil remediation. These fungi are likely to be most beneficial in diverse nutrient deficient ecosystems where the proportion of plants able to form Mycorrizhas is high. Benefits from use of the fungi include better plant growth by increasing nutrient uptake, efficient recycling of nutrients, and stabilization of the soil in the long term.

Rehabilitation trials are on with Mycorrhizal inoculations for a timeframe of eight years at the field station at Gualpahari (Haryana), to yield significant improvement in biomass and nutritional status, while eliminating use of chemical fertilizer completely. Another outlook on reclamation is to use fly ash as a special kind of soil. However, the ash characteristics, particularly, its heavy metal composition, warrants the selection of plant species best suited for reclamation.

WAY FORWARD

The degradation of our rural environment is palpable, albeit gradual. Even though governments acknowledge the gravity of environmental concerns, they have tended to divorce economic growth targets from environmental impact assessment. There is a need to change the policy focus which, so far, has been on a curative (if such a thing is possible) approach to environmental problems and hazards, to a preventive approach. Such an approach would include some of the following measures. For better water management infrastructure there has to be a comprehensive policy integrating water planning with overall economic and social planning. The pursuit of particular goals in water development should also aim at strengthening the objectives in the management of other resources. This implies developing multipurpose projects with possibilities of adding hydroelectric power production and development of irrigation. Second, the economic value of water must be recognized. Water should be a priced commodity. Third, local people need to be consulted on goals and strategies in water planning. This is important for rural areas and participation must take place from the first stages of decision-making. Fourth, impact assessment should be a must for any water development programmes.

This can bring to attention the threats to water quality as also to human welfare and social stability. This will also help unite the science of environmental analysis with the politics of resource management. In particular, we need to pay attention to the regeneration and revival of old irrigation systems and projects. In the dry land areas of the country, appropriate watershed development is critical.

We need to bring wastelands and degraded lands into productive use, either under crops or under agro-forestry, and to improve credit flows to our farmers through innovative methods. Technological interventions are essential to improve agricultural productivity and to widen the range of products. Institutional structures governing rural activities also need to be reformed and strengthened. It is necessary to evolve a new approach to agricultural policy based on a careful assessment of current constraints and possibilities. Both land and water are crucial constraints for expanding production in agriculture. The extent of forest cover has declined alarmingly. Thus, there is little possibility of increase in the cultivated area in the country, and indeed perhaps an eventual decline as urban demand and environmental imperatives lead to conversion of some agricultural land. There is, therefore, no alternative but to focus on raising the productivity of our land in a sustainable manner.

For this there is a need to develop and disseminate agricultural technologies. Over the years India has developed an extensive system of agricultural research centres and extension services. However, their efficacy is in question. Strengthening the agricultural research and development system and a significant improvement in the sophistication of the technology dissemination methodologies are essential to achieving rapid and sustained growth in agricultural productivity. An increased focus on subsistence crops and technologies in rainfed/dry land areas would be appropriate in this regard. Every effort needs to be made to bring presently uncultivated land into productive use, whether in agriculture or in forestry. For this, it will be essential to evolve a comprehensive land-use policy which will lay out the contours of the ownership and institutional framework that will encourage the productive utilization of such land.

Potential policy options for agricultural watersheds would include controlling sources, and thereby minimizing nutrient and waste input, while processing and moving surplus manure to nutrient-poor areas. Regulation of animal stock densities has to be ensured. On-farm nutrient budgets could be promoted by public institutions and community involvement. Finally, educating the people is important, and has to be done with the involvement of various stakeholders.

❐

6

Survey of Natural Resources

Environmental problems and issues received special attention of the Government of India during the beginning of the Fourth Five Year Plan. As a follow-up step, a National Committee of Environmental Planning and Co-ordination (NCEPC) was set up in 1972 under the Department of Science and Technology. A separate Empowered Committee was set up in 1980 for reviewing the existing legislative measures and administrative machinery for ensuring environmental protection and for recommending ways to strengthen them.

On the recommendations of this Empowered Committee, a separate Department of Environment was set up in 1980 which was subsequently upgraded to a full-fledged Ministry of Environment and Forests in 1985 to serve as the focal point in the administrative structure of the Government of India for the planning, promotion and co-ordination of environmental and forestry programmes. The State departments of environment, Central and State Pollution Control Boards, the Botanical and Zoological Survey of India, the Forest Survey of India, the National River Conservation Authority (formerly Central Ganga Authority), the National Afforestation and Ecodevelopment Board, the Indian Council for Forestry Research and Education, the Wildlife Institute of India, the National Museum for Natural History, etc. , are the Ministry's partners in carrying out environmental protection activities.

SURVEY OF NATURAL RESOURCES

The Botanical Survey of India (BSI), established in 1980, is responsible for surveying and identifying the plant resources of the country. With its headquarters at Calcutta and nine circles located in different regions of the country, the BSI undertakes exploration tours of the country regularly and the results of such tours are published in the form of national, state and district flora.

Established in 1916, the Zoological Survey of India (ZSI) is responsible for carrying out surveys of the faunal resources of the country. While the headquarters of the Zoological Survey of India is at Calcutta, it has 16 regional stations located in different parts of the country. ZSI also undertakes regular faunistic survey tours of the entire country.

The Forest Survey of India (FSI), established in 1981,is entrusted with the task of surveying the forest resources of the country. Besides the headquarters at Dehra Dun, FSI has four regional offices located at Bangalore, Calcutta, Nagpur and Shimla. It prepares thematic maps on 1: 50,000 scale and forest vegetation maps on 1: 2,50,000 scale of the country.

The thematic maps are prepared for the entire country on a ten-year cycle. The vegetation maps are prepared on a two-year cycle for the entire country. The State of Forest Report, 1997 is the sixth report in the series of biennial reports published since 1987 and presents a comprehensive account of forest cover.

NATIONAL CONSERVATION STRATEGY

The National Conservation Strategy and Policy Statement on Environment and Development, adopted by the Government of India in June 1992, lays down strategies and actions for integration of environmental considerations in the development activities of various sectors of the country, thus paving the way for achieving sustainable development. The action points pertaining to individual ministries/departments ensure that they take action for reorienting their policies and programmes in conformity with the strategy.

BIOSPHERE RESERVES

Biosphere reserves are multi-purpose protected areas to preserve the genetic diversity in representative eco-systems.

The major objectives of biosphere reserves are:

- To conserve diversity and integrity of plants, animals and micro-organisms;
- To promote research on ecological conservation and other environmental aspects
- To provide facilities for education, awareness and training.

So far eleven biosphere reserves have been set up.

They are:

- Nilgiri
- Nanda Devi
- Nokrek
- Great Nicobar
- Gulf of Mannar
- Manas
- Sunderbans
- Similipal
- Dibru Saikhowa
- Dehong Deband
- Pachmarhi.

Comprehensive guidelines for them emphasize formulation of eco-development and demonstration projects, development of data-base conservation plans of key species, establishment of research stations and implementation of social welfare activities. Non-governmental organisationsare being involved in the biosphererereserve programmes for creation of publicawareness.

WETLANDS, MANGROVES AND CORAL REEFS

India has a wealth of wetland eco-systems distributed in different geographical regions from the cold arid zone of Ladakh in the North to the wet humid climate of Imphal in the East, the warm arid zone of Rajasthan in the West to the tropical monsoonic Central India and the wet and humid zone of Southern Peninsula. Most of the wetlands in India are directly or indirectly linked with major river systems such as Ganga, Brahmaputra, Narmada, Tapti, Godavari, Krishna, Cauveri, etc.

A National Level Committee constituted to advise the Government on appropriate policies and measures to be taken for conservation and management of the wetlands, has so far identified 20 wetlands for conservation and management on priority basis. Steering Committees have been set up by the concerned State governments in which representatives of State government departments, universities and research institutions are included. Nodal research/academic institutions have been identified for each of the selected wetlands.

Management Action Plans have been drawn up for most of the identified wetlands. A directory on wetlands in India has been published which gives information on location, area and ecological categorisation of wetlands in different parts of the country. India is a signatory to the Convention on Wetlands of international importance, especially as Waterfowl Habitat (Ramsar Convention) and six Indian Wetlands, *viz.*,

Keoladeo National Park, Bharatpur and Sambar (Rajasthan), Chilka (Orissa), Loktak (Manipur), Wullar (Jammu and Kashmir), and Harike (Punjab) have been designated under this Convention. Mangroves are salt-tolerant forest ecosystems found mainly in the tropical and sub-tropical inter-tidal regions of the world. They are reservoirs of a large number of plant and animal species associated together over a long evolutionary period and exhibiting remarkable capacity for salt tolerance. They stabilize the shoreline and act as a bulwark against encroachments by the sea. India harbours some of the best mangroves in the world and these occur all along the Indian coastline in sheltered estuary, tidal creeks, backwaters, salt marshes and mud flats covering a total area of 4,827 sq km.

Under the scheme on Conservation and Management of Mangroves, 15 mangrove areas have been identified for intensive conservation and management purposes: Northern Andaman and Nicobar, Sunderbans (West Bengal), Bhitarkanika (Orissa), Coringa, Godavari Delta and Krishna Estuary (Andhra Pradesh), Mahanadi Delta (Orissa), Pitchavaram and Point Calimer (Tamil Nadu), Goa, Gulf of Kutch (Gujarat), Coondapur (Karnataka), Achra/ Ratnagiri (Maharashtra) and Vembanad (Kerala). Management action plans for all the 15 mangrove areas have been sanctioned. Coral reefs are shallow- water tropical marine ecosystems, characterised by high biomass production and rich floral and faunal diversity. Four coral areas, Gulf of Mannar, Andaman and Nicobar Islands, Lakshadweep Islands and Gulf of Kuchch have been identified for conservation and management. State-level steering committees have been constituted for the formulation and implementation of management action plans.

Such action plans have been sanctioned for Andaman and Nicobar and Gulf of Mannar coral reefs so far. The objectives of the Convention on Biological Diversity (CBD) are: the conservation of Biological Diversity; the sustainable use of its component and the fair and equitable sharing of the benefits arising out of the utilization of genetic resources. Following the ratification of the Convention on Biological Diversity by India in 1994, several steps have been initiated both to meet the commitments under the Convention and to realise the opportunities offered by the CBD. A draft National Action Plan on Biodiversity has been finalised which seeks to consolidate the on-going efforts of conservation and sustainable use of biological diversity and to establish a policy and programmes regime for the purpose. A legislation on bio-diversity is also under finalisation. India regularly participates in important international conventions on Biological Diversity.

FORESTS

Forests are a renewable source and contribute substantially to economic development. They play a major role in enhancing the quality of environment. The country has an area of 633.4 lakh hectares notified as forests which represents 19.27 per cent of geographic area. Out of this, dense forest (crown density more than 40 per cent) accounts for 11 per cent, open forest (crown density 10-40 per cent) represents eight per cent, while mangrove forest occupies 0.15 per cent.

FOREST POLICY AND LAW

India is one of the few countries which have a forest policy since 1894. It was revised in 1952 and again in 1988. The main plank of the Forest Policy of 1988 is protection, conservation and development of forest.

Its aims are:

- Maintenance of environmental stability through preservation and restoration of ecological balance;
- Conservation of natural heritage;
- Check on soil erosion and denudation in catchment area of rivers, lakes and reservoirs;
- Check on extension of sand dunes in desert areas of Rajasthan and along coastal tracts;
- Substantial increase in forest tree cover through massive afforestation and social forestry programmes;
- Steps to meet requirements of fuelwood, fodder, minor forest produce and soil timber of rural and tribal populations;
- Increase in productivity of forest to meet the national needs;
- Encouragement of efficient utilisation of forest produce and optimum substitution of wood
- Steps to create massive people's movement with involvement of women to achieve the objectives and minimise pressure on existing forests.

The entire gamut of forest activities are being given a new orientation inthe light of the National Forest Policy of 1988. In order to operationalise the National Forest Policy 1988, a National Forestry Action Programmes (NFAPis being prepared. As a part of this exercise State Forestry Action Programmes are also being prepared for each State.

Under the provisions of the Forest (Conservation) Act, 1980,prior approval of the Central government is required for diversion of forest lands for non- forest purposes. Since the enactment of the Act, the rate of diversion of forest land has come down to around 25,000 hectares per annum from 1.43 lakh hectares per annum, before1980.During 1998, 851 proposals from various State and UT governments were processed under this Act. A scheme titled 'Association of Scheduled Tribes and Rural Poor in Regeneration of Degraded Forests on Usufruct Sharing Basis' is under implementation in nine States of the country. Besides improving the forest cover, the scheme also aims at providing wage employment and usufructs to the tribal people. Joint Forest Management (JFM) is being practised in 21 States of the country. About 7 million hectares of degraded forests in the country are being managed and protected through approximately 35,000 village Forest Protection Committees.

FOREST FIRE CONTROL

Fire is one of the major factors responsible for destruction of forests in the country. In India, most of the fires are man-made and are generally deliberate and on rare occasions accidental. Some of the reasons for intentional fire are grazing, *mahua* seeds and flower collection, *tendu* leaf collection, poaching, shifting cultivation, etc.

In order to reduce the incidents of fire in forests, the following steps have been undertaken by the forest department:

- Development of firelines
- Establishment of watchtowers, besides employing firewatchers during fire seasons.

The UNDP-assisted Modern Forest Fire Control Project, initiated in 1984 in Chandrapur (Maharastra) and Haldwani/Nainital (Uttar Pradesh), is in operation in 11 States of the country. Financial assistance is provided to the state governments for hand tools, fire-resistant clothing, wireless communication sets, fire-fighting instruments, construction of watch-towers, creation of fire lines, etc. The project is being revised to include the monitoring of forest fire incidents in the country.

WILDLIFE

The National Wildlife Action Plan provides the framework of strategy as well as programmes for conservation of wildlife. The Indian Board of Wildlife, headed by the Prime Minister, is the apex advisory body overseeing and guiding the implementation of various schemes for wildlife conservation. At present the protected area network comprises 84 national parks and 447 sanctuaries covering 4.5 per cent of the total geographical area of the country.

The Wildlife (Protection) Act, 1972 adopted by all states except Jammu and Kashmir (which has its own Act), governs wildlife conservation and protection of endangered species. The Act prohibits trade in rare and endangered species. An Inter-State Committee has been set up to review the Wildlife (Protection) Act, 1972 and other laws. India is a signatory to the Convention on International Trade in Endangered Species of wild flora and fauna (CITES). Under this, export or import of endangered species and their products is subject to strict control. India is also a signatory to the

Memorandum of Understanding concerning the conservation of Siberian Cranes. The Centre provides financial and technical assistance to states for development and improved management of national parks; protection of wildlife and control of poaching and illegal trade in wildlife products; eco-development in areas around national parks and sanctuaries; conservation of elephant and its habitat; and conservation of rhinos in Assam.

Under Project Tiger, launched in April 1973, 25 Tiger Reserves have been set up in 14 States covering an area of about 33,875 sq km. The tiger population now reported by 14 States is 3,435 only. Financial assistance is available for voluntary relocation of tribal families from the Protected Areas. Under the Project Elephant, States having free-ranging population of wild elephants are being given financial as well as technical and scientific assistance to ensure long-term survival of identified viable population of elephants in their natural habitats. A Central Zoo Authority has been set up to look after the management of zoological parks in the country. It coordinates the activities of over 200 existing zoos and also supervises the exchange of animals on a scientific basis.

A national policy on zoos prepared by the Authority has been adopted which provides appropriate directions to the Government and other zoo operators. The export of the shed antlers of deer (Chital and Sambhar) and articles made of these have been banned. Export of 29 medicinal plants, parts thereof, and pure extracts of these plants have been banned.

ANIMAL WELFARE BOARD

The Animal Welfare Board of India, established in 1962 under the provisions of the Prevention of Cruelty to Animals Act, 1960 is an autonomous organisation of the Ministry of Environment and Forests working for the cause of animal welfare in the country.

More than 9,000 Honorary Animal Welfare Officers have been appointed by the Board so far to check cruelty to animals and to protect them.

A series of camps for training the members of the Society for Prevention of Cruelty to Animals and animal welfare organisations are being held all over the country. Animal Welfare Fortnight is celebrated from 14 January every year.

ENVIRONMENTAL IMPACT ASSESSMENT

Impact assessment is a pointer to the environmental compatibility of the projects in terms of their location, suitability of technology, efficiency in resource utilisation, recycling and so on.

Impact assessment was introduced in India in 1978 and now covers projects such as:

- – River valley;
  – Thermal power;
  – Mining;
  – Industries;
  – Atomic power;
  – Rail, road, highways, bridges;
  – Ports and harbours;
  – Airports;
  – New towns and
  – Communication projects
- Those which require the approval of the Public Investment Board/Planning Commission/Central Electricity Authority.
- Those referred to the Ministry of Environment and Forests by other ministries.
- Those which are sensitive and located in environmentally degraded areas.
- Public sector undertakings of the Centre where the project cost is more than `50 crore.

A notification issued in January 1994 makes Environmental Impact Assessment statutory for 29 categories of developmental projects under various sectors such as industrial, mining, irrigation, power, transport, tourism, communication, etc. The Environmental Impact Assessment (EIA) Notification was amended in 1997 to provide for public hearing as well as for empowering State governments for according environmental clearance in respect of certain Thermal Power Projects. As per the provisions of the EIA Notification 1994, mining projects are subject to environmental clearance.

Until now, all proposals for prospecting and exploration were required to obtain the clearance. However, it has been now decided to exempt prospecting through aerial survey and/or test drilling in smaller areas from the clearances. Applications complete in every material aspect are normally examined and decision conveyed to the applicants within 30 days in cases of site clearance and 120 days in cases of environmental clearance of projects. Once an application has been submitted by the project authority, the preliminary scrutiny of the project is done by the respective technical divisions and the overall appraisal of the projects is undertaken by specially constituted environmental appraisal committees of experts.

In addition, special groups/committees and task forces are constituted as and when needed for expert inputs on major projects. After detailed scrutiny and assessment, the appraisal committee makes its recommendations for approval or rejection of the project. To ensure transparency, the position of forest and environmental clearance has been brought out on web site: httpc/ www. nic. in/envfor since February 1999.Depending on the nature of the project, certain safeguards are recommended. For monitoring and timely implementation of safeguards suggested, six regional offices of the Ministry have been set up at Shillong, Bhubaneswar, Chandigarh, Bangalore, Lucknow and Bhopal.

AUTHORITIES UNDER ENVIRONMENT PROTECTION ACT, 1986

A National Environmental Appellate Authority has been constituted to hear appeals with respect to rejection of proposals from the environmental angle. The objective is to bring in transparency in the process and accountability, and to ensure the smooth and expeditious implementation of developmental schemes and projects.

An Environmental Impact Assessment Authority for the National Capital Region has been constituted to deal with environmental protection problems arising out of projects planned in the National Capital Region (NCR). An Aquaculture Authority has been constituted to deal with the situation created by the Shrimp Culture industry in the coastal States and Union Territories.

The Central Ground Water Authority for regulation and control of ground water management has initiated action regarding registrations for ground water pollution/depletion. It has also initiated mass awareness programmes. Besides this, different authorities have been created for dealing with specific problems in the States of Tamil Nadu and Maharashtra.

PREVENTION AND CONTROL OF POLLUTION

The policy statement on Abatement of Pollution, adopted in 1992, provides instruments in the form of legislation and regulation, fiscal incentives, voluntary agreements, educational programmes and information campaigns to prevent and control pollution of water, air and land. Since the adoption of the policy statement, the focus of activities has been on issues such as promotion of clean and low waste technologies, waste water minimisation, reuse/ recycling, improvement of water quality, environment audit, natural resource accounting, development of mass-based standards, institutional and human resource development, etc.

The whole issue of pollution prevention and control is dealt with by a combination of command and control methods as well as voluntary and regulations, fiscal measures, promotion of awareness and involvement of public. To facilitate industries in preparing environmental statements, sectorspecific environmental audit manuals have been prepared.

A software package, *Paryavaran,* along with a user manual has been prepared for analysis of information submitted in these environmental statements and distributed to all the State Pollution Control Boards. Training programmes are also being organised for officials of CPCB and SPCB to enable them to use the software. An "Eco-mark" label has been introduced to label consumer products that are environment-friendly.

Nineteen products have been identified for labelling and 18 notifications have been issued so far on different products criteria. The Bureau of India Standards (BIS)/Directorate of Marketing and Inspection (DMI) is the implementing agency for this scheme. So far one license has been granted by the BIS to a product under the soaps category. Under the scheme for adoption of clean technology in small-scale industries and for extending necessary technical support, training and awareness programmes for personnel in Small Industry Development Organisation and for entrepreneurs are being organised.

'From Waste to Profits', a manual giving guidelines for waste minimisation, has been prepared. Sector-specific manuals on waste minimisation in the areas of pulp and paper, pesticides formulations and textiles, dyeing and printing and electroplating, have also been prepared. Waste Minimisation Circles (WMCs) are being established to promote group efforts in increasing productivity and improving the environmental conditions in small and medium-scale industries through adoption of waste minimisation techniques. Fifteen Waste Minimisation Circles have been established so far in different industrial clusters across the country. An Indian Centre for the Promotion of Cleaner Technologies (ICPC) with a network of institutions including industries, academic institutions and other user agencies is being set up for which the World Bank has provided $2 million as grant-in-aid.

The Centre will provide evaluated and ranked technology options to entrepreneurs. A major programmes for improvement of automotive fuels such as motor gasoline and diesel is being introduced in the country in a phased manner, with the objective of improving the air quality of the cities of our country. Low-lead petrol (0.15 gms/litre lead content) has been introduced at all retail outlets of the country from 1 April 1996. Unleaded petrol has been made available in selected retail outlets of the four metro cities of Delhi, Mumbai, Calcutta and Chennai from 1 April 1995. From 1 June 1998 unleaded petrol is also available in all state capitals and major towns. From 1 September 1998 only unleaded petrol has been available at all the outlets in Delhi and it is extended to National Capital Territory of Delhi from 1 January 1999.

Ban on 2T oil has been imposed in the National Capital Territory of Delhi from 31 December 1998. High-speed diesel containing 0.5 per cent sulphur has been introduced in the four metro cities from 1 April 1996. HSD with 0.25 per cent sulphur is being supplied in the Taj Trapezium area from 1 September 1996. It is also being supplied in Delhi from 15 August 1997. Stricter emission norms for new vehicles effective from 1 April 2000 have been notified.

CENTRAL POLLUTION CONTROL BOARD

The Central Pollution Control Board (CPCB) is the national apex body for assessment, monitoring and control of water and air pollution. The executive responsibilities for enforcement of the Acts for Prevention and Control of Pollution of Water (1974) and Air (1981) and also of the Water (Cess) Act, 1977 are carried out through the Board. The CPCB advises the Central Government in all matters concerning the prevention and control of air, water and noise pollution and provides technical services to the Ministry for implementing the provisions of the Environment (Protection) Act, 1986.

Under this Act, effluent and emission standards in respect of 61 categories of industries have been notified. Seventeen categories of heavily polluting industries have been identified. They are: cement, thermal power plant, distilleries, sugar, fertilizer, integrated iron and steel, oil refineries, pulp and paper, petrochemicals, pesticides, tanneries, basic drugs and pharmaceuticals, dye and dye intermediates, caustic soda, zinc smelter, copper smelter and aluminium smelter. Out of a total of 1,551 units identified under these 17 categories, 1,266 units have installed adequate facilities for pollution control and 130 units have been closed down. The Central Pollution Control Board, in consultation with State Pollution Control Boards, has identified critically polluted areas in the country which need special attention for control of pollution.

These are:

- Vapi (Gujarat)
- Singrauli (Uttar Pradesh)
- Korba, Ratlam, Nagda (Madhya Pradesh)
- Digboi (Assam)
- Bhadravati (Karnataka)
- Howrah (West Bengal)
- Manali and North Arcot (Tamil Nadu)
- Visakhapatnam and Patancheru, (AndhraPradesh)
- Chembur (Maharashtra), Najafgarh (Delhi)
- Govindgarh (Punjab)
- Udyog Mandal (Kerala)
- Parwanoo and Kala Amb (Himachal Pradesh)

The CPCB in collaboration with the SPCBs monitors the quality of fresh water resources of the country through a network of 480 monitoring stations located all over the country. Based on such monitoring, 13 heavily polluted and 26 medium-polluted river stretches have been identified. Under the National Ambient Air Quality Monitoring programmes, 290 stations covering over 90 towns/cities monitor the ambient air quality of the country.

The Central and State Pollution Control Boards regularly conduct surveys in different cities of the country pertaining to vehicular and noise pollution, sanitation status, status of solid waste, etc. A survey on the status of solid waste conducted in 299 Class I cities of the country indicates that 62 per cent of the total solid waste generated in the country comes from the 23 metro cities of the country. The average per capita generation of solid waste for Class I cities is about 376 gms per person per day. A total of 1,532 grossly polluting industries in 24 States/Union Territories have been identified under the National River Action Plan. Comprehensive River Basin Documents for the rivers Ulhas, Brahmaputra, Pennar, Indus Part II, Rishkulya and Chaliyar are being prepared by the Board. The Central Pollution Control Board has a NGO Cell for interacting with NGOs.

Simple water-testing kits are distributed free of cost to selected NGOs and financial assistance provided to them for conducting mass awareness programmes relating to prevention and control of pollution. The White Paper on status of pollution in Delhi with an Action Plan for its control prepared earlier is being implemented. The Action Plan contains specific measures for control of pollution relating to vehicular pollution, water pollution, industrial air pollution, solid waste, hospital wastes, industrial hazardous wastes, noise pollution and people's participation in making Delhi a cleaner city. Directions have been issued by the National Capital Territory of Delhi for imposing restrictions on all commercial vehicles in Delhi in a time-bound programmes beginning from April 1998. Based on the recommendations of a National Level Committee on Noise Pollution, directions have been issued to the State governments to check noise pollution from bursting of crackers. Apart from introducing unleaded petrol in the National Capital Territory of Delhi from 1 September 1998 and the National Capital Region from 1 January 1999, there is a proposal to promote the use of CNG in buses, private vehicles and taxies.

There is also a proposal to introduce a certain percentage of ethanol/ methanol in gasoline so as to reduce the emission of carbonmonoxide and hydrocarbons. It was also proposed to set up a high level body to look into issues concerning vehicle emissions. There is also a plan to introduce higher parking fees, to augment parking spaces in the city of Delhi and to introduce no traffic zones.

MANAGEMENT OF HAZARDOUS SUBSTANCES

The Ministry of Environment and Forests is the nodal agency for the management and control of hazardous substances which include hazardous chemicals, waste and micro-organisms.

The following rules have been notified under the Environment (Protection) Act, 1986:

- Manufacture, Storage and Import of Hazardous Chemicals, 1989;
- Hazardous Wastes (Management and Handling) Rules, 1989;
- Manufacture, Use, Import, Export and Storage of Hazardous Micro-organisms/Genetically Engineered Organisms or Cell, 1989
- Biomedical Waste Rules, 1998.

A central control room has been set up in the Ministry to deal with emergencies arising from hazardous chemicals and a Crisis Alert System has been established. Guidelines for preparation of crisis management plans have been issued to the state governments and financial support is being provided to them to strengthen infrastructure for the purpose. Emergency Response Centres have been set up at Bhopal, Baroda, Manali and Khapali

The Red Book entitled 'Central Crisis Group Alert System', which includes names, addresses and telephone numbers of the Central and state authorities and experts to be contacted in case of emergency, has been updated and circulated to all concerned. A sub-scheme entitled 'Industrial Pocket-wise Hazard Analysis' has been in operation since the Eighth Five Year Plan.

Hazard analysis studies of 37 industrial pockets have since been completed. The remaining 8 Hazard analysis studies are being offered to the consultants. A Public Liability Insurance Act has been enacted to provide immediate relief to the victims of accidents by hazardous chemical industries. India is a signatory to the UNEP-sponsored Convention on Control of Transboundary Movement of Hazardous Wastes which was adopted at Basel, Switzerland, by 126 governments of the world in 1989. The Convention aims at checking the reported illegal traffic in hazardous wastes from one country to another. A scheme on National Register of Potentially Toxic Chemicals (NRPTC) has been started for setting up the basic infrastructure for implementing the London guidelines for the exchange of information on chemicals in international trade, including the procedure for prior informed consent. Nine Regional Registers (RRPTC) have been set up in Uttar Pradesh, Madhya Pradesh, Andhra Pradesh, Gujarat, Orissa, Kerala, Himachal Pradesh, Punjab and Maharashtra.

A National Poison Information Centre has been set up at the Department of Pharmacology in the All India Institute of Medical Sciences, New Delhi. An antidote data-bank is being developed at the Centre. The Ministry has constituted a National Waste Management Council to suggest ways and means for effective utilisation of wastes generated in the country. Three sub-groups have been set up by the Council to deal with the major categories of wastes, *viz.*, industrial, urban and rural.

The sub-groups have been entrusted with the task of identifying wastes, suggesting technological action points, including legislation, taxes and incentives. A Pilot Project on Municipal Solid Waste Management has been sanctioned to the Hyderabad Municipal Corporation. Special emphasis is being laid on promoting the use of flyash in various fields, and the State governments have been asked to prepare a perspective action plan for 50 per cent utilisation of flyash by the turn of the century. A high-powered committee on hazardous wastes has been constituted to oversee the strict implementation of the rules and regulations on management of these wastes.

For the disposal of hazardous wastes, 15 sites have been identified in the country for preparation of secured land fills. Erring industries of the village Bichchri, Rajasthan, have been ordered to undertake remedial work on the 320-hectare land contaminated due to indiscriminate disposal of hazardous wastes. The Hazardous Wastes (Management and Handling) Rules, 1989 and Manufacture, Storage and Import of Hazardous Chemical Rules, 1989 have been amended. Final Rules on Bio-medical Wastes have been notified. A new set of rules, 'Emergency Planning Preparedness and Response to Chemical Accidents', has been notified. A notification has been issued banning import of wastes containing Beryllium, Selenium, Chromium (Hexavalent), Thallium, pesticides, herbicides and insecticides as well as Waste Asbestos (dust and fibre), and waste containing PCB, PCT and PBB.

NATIONAL RIVER CONSERVATION AUTHORITY

The Central Ganga Authority (CGA) established in 1985, lays down the policies for works to be taken up under the Ganga Action Plan (GAP). With the approval of the National River Conservation Plan (NRCP) in 1995, the CGA has been reconstituted as the National River Conservation Authority (NRCA) and the Ganga Project Directorate has been redesignated as the National River Conservation Directorate (NRCD).

The NRCA is assisted by a Steering Committee and a Monitoring Committee. The NRCD of the Ministry services these committees and coordinates the implementation of the schemes under the Ganga and other Action Plans. The main component of the Ganga Action Plan is diversion of sewage flowing into the river to other locations for treatment and conversion into valuable energy sources. As against the target of creating infrastructure capable of intercepting, diverting and treating 873 million litres per day (mld) of domestic waste, infrastructure capable of treating 728 mld of municipal sewage has been created.

Out of 261 schemes sanctioned, 254 have been completed and almost all low-cost sanitation schemes and construction of electric crematoria have been completed, leading to significant reduction in the pollution load of the river Ganga. The Ganga Action Plan, Phase II has been merged with the National River Conservation Plan (NRCP) through a Government Resolution. The expanded NRCP now covers 141 towns located along 22 inter-state rivers in 14 States.

The total cost of the schemes is `2,014.39 crore. Pollution abatement works in 46 towns located along 18 inter-state rivers have been taken up. Works have also been taken up at 59 towns on the Main Stem of River Ganga. The scheme of cleaning the river Yamuna covers 21 towns in three states and its estimated cost is `509.45 crore.

Under the Gomti river scheme, works are proposed in three towns at an estimated cost of `64 crore. The scheme of pollution abatement of the river Damodar was approved in October 1996 at a cost of `24.54 crore. Works are proposed in 12 towns in two States. A National Lake Conservation Plan has also been initiated under which 11 urban lakes have been identified for immediate conservation. The World Bank has agreed to provide financial assistance for Conservation and Management of Lakes and Wetlands in Rajasthan and Punjab.

NATIONAL AFFORESTATION AND ECO-DEVELOPMENT BOARD

The National Wastelands Development Board (NWDB) established in May 1985, was bifurcated into a new Department of Wastelands Development and a National Afforestation and Eco-Development Board (NAEB) in 1992.While the NAEB is a part of the Ministry of Environment and Forests, the Department of Wastelands Development has been transferred to the Ministry of Rural Development.

The responsibilities of the NAEB include promotion of afforestation, tree plantation, ecological restoration and eco-development activities in the country with special attention to the degraded forest areas and lands adjoining the forest areas, national parks, sanctuaries and other protected areas, as well as the ecologically fragile areas like the Western Himalayas, Aravallis, Western Ghats, etc. The NAEB seeks to carry out its responsibilities through several schemes such as the integrated afforestation and eco-development project, area-oriented fuelwood and fodder projects, non-timber forest produce (including medicinal plants) scheme, aerial seeding, seed development, technology extension, grants-in-aid to voluntary agencies, etc.

During the Ninth Five Year Plan, four key elements, *viz.*, Promotion of Joint Forest Management, Micro planning, Incorporation of improved technologies, and Monitoring and Evaluation of projects, will be emphasized while implementing these schemes. Eco-task forces of exservicemen have been deployed in the states of Uttar Pradesh, Rajasthan and Jammu and Kashmir for undertaking eco-restoration work through afforestation, pasture development and other water and soil conservation measures in selected highly degraded and inhospitable areas of these states.

To commemorate the 50th Anniversary of India's Independence, the Board provided financial assistance to State Forest Departments for distributing seedlings to all Village Panchayats to enable every village to raise a cluster of 50 trees. Wastelands maps on 1: 50,000 scale have been prepared for 237 districts of the country using satellite data. The NAEB has seven Regional Centres located at different universities and national-level institutions of the country. These centres help the NAEB in promoting extension of replicable technologies and in disseminating research findings to a wider audience. These centres also carry out problemspecific studies and undertake evaluation of NAEB's programmes in the field.

RESEARCH PROMOTION

ENVIRONMENTAL RESEARCH PROGRAMMES

The environmental research programmes aims at developing strategies for sustainable resource management and creating facilities and trained manpower to conduct research in priority areas.

Research projects are supported under three main schemes:

- Man and Biosphere Programmes;
- Environment Research Programmes;
- Action-oriented Research Programmes for Eastern and Western Ghats.

Many small, specific research projects are also being supported. Expert Committees and Expert Working Groups make recommendations for undertaking research in priority areas, scrutinise proposals, evaluate progress and recommend suitable mechanisms for implementation of research projects. Priority areas of research have been identified in five fundamental elements, *viz.*, land, water, air, space and energy. During the year, 45 new projects were sanctioned, 45 were completed and 112 projects were serviced.The Framework Convention in Climate Change ratified by 127 countries of the world was ratified by India during 1993.

India participated in the 3rd Conference of Parties to this Convention held in Kyoto, Japan in December 1997 during which a protocol was agreed upon which calls for an average cut in the greenhouse gas emission, of 5.2 per cent below the 1990 levels, to be achieved between 2008 and 2012 AD. Specific research projects are being sponsored by the Ministry in the area of climate change for taking precautionary measures to anticipate, prevent and minimise the causes of climate change and mitigate its adverse effects. The Asia Least Environment Facility has been completed.

GB PANT INSTITUTE OF HIMALAYAN ENVIRONMENT AND DEVELOPMENT

The GB Pant Institute of Himalayan Environment, established in 1988, is an autonomous organisation of the Ministry of Environment and Forests for developing strategies, technologies and knowledge base for ecologically sound development of the Himalayan region. Besides undertaking actionoriented research projects, the Institute has also installed three weathermonitoring stations for collecting and analysing climatic data of the Himalayan region.

Units of the Institute are located in Uttar Pradesh, Sikkim, Nagaland and Himachal Pradesh. The Ministry also sanctions specific research projects under various programmes such as Biosphere Reserves, Wetlands, Mangroves and Coral Reefs, etc. The research activities of the National River Action Plan focus on identifying economically viable and technically feasible solutions for controlling the microbial pollution in the rivers. During the current year, it is proposed to initiate research studies in estimation of non-point pollution, heavy-metals uptake in vegetables and river fauna, etc.

FORESTRY RESEARCH

Forestry research is the primary responsibility of the Indian Council of Forestry Research and Education located at Dehra Dun. The following forestry research institutes and centres under the Council are responsible for undertaking research in their respective eco-climatic zones:

- Forest Research Institute, Dehra Dun
- Institute of Arid Zone Forestry Research, Jodhpur
- Institute of Rain and Moist Deciduous Forests, Jorhat
- Institute of Wood Sciences and Technology, Bangalore
- Tropical Forestry Research Institute, Jabalpur
- Institute of Social Forestry and Eco-rehabilitation, Allahabad
- Instt. of Forestry Research and Human Resources Development, Chhindwara
- Advanced Centre for Bio-technology and Mangrove Forests,Hyderabad

In addition, the Indian Plywood Industries Research and Training Institute, Bangalore, has been transferred to the Ministry of Environment andForests from the Ministry of Industry.

Besides forestry research in wood panel products, this Institute imparts training to workers in optimum utilisation of timber. The Indian Institute of Forest Management, Bhopal, also undertakes education, training, research and consultancy in forest management.

WILDLIFE RESEARCH

Research programmes in wildlife are carried out by the Wildlife Institute of India, Dehra Dun, and the Salim Ali Centre for Ornithology and Natural History, Coimbatore. Several projects on habitat evolution, elephant movement, ecology of *gharials* and turtles, status of endangered species, behavioural ecology, bio-diversity, resource study conservation, ecology and management of specific animals, etc. , are being carried out by both these Institutes.

NATIONAL NATURAL RESOURCES MANAGEMENT SYSTEM

The basic objective of the Natural Resources Management System (NRMS) is the utilisation of Remote Sensing Technology with the conventional methods for optimal use and management of the natural resources of the country. A Standing Committee on Bio-resources and Environment advises the Ministry on various aspects relating to the implementation of this scheme. Specific research projects are also supported under this scheme. Two Sub-Committees, the Technical and Financial Sub-Committee and the Forest Sub-Committee, help the Standing Committee in its task.

EDUCATION, AWARENESS AND INFORMATION

Priority is accorded by the Ministry of Environment and Forests to promote environmental education, create environmental awareness among various age-groups and to disseminate information through Environmental Information System (ENVIS) network to all concerned. A major initiative to include environment education as a separate and compulsory subject in the educational curricula has been taken by the Ministry at all levels of formal education, *i.e.*, secondary, senior secondary and tertiary levels.

A discussion paper prepared on strengthening environment education was presented by the Minister for Environment and Forests at the State Education Ministers' Conference held from 22 to 24 October 1998. The paper was adopted by the Conference. The Chief Ministers were urged to introduce environment education in the school curricula from the 1999-2000 academic session.

Maharashtra is the first State to introduce the subject in the school curriculum. Special emphasis is given to non-formal environmental education through seminars/symposia/workshops, training programmes, eco-camps, audiovisual shows, etc. The Ministry has been organising a National Environmental Awareness Campaign (NEAC) since July 1986. As a part of this campaign, 19 November to 18 December every year is observed as the National Environment Month. The main theme for the 1998 campaign was 'Keep Our Water Resources Clean'. A large number of organisations have been granted financial assistance by the Ministry to organise various activities for creating environmental awareness. The Ministry also provides financial assistance for setting up eco-clubs in schools.

A new scheme, *Paryavaran Vahini,* was launched in 1992-93 to create environmental awareness and to ensure active public participation by involving the local people in activities relating to environmental protection. *Paryavaran Vahinis* are proposed to be constituted in 194 selected districts all over the country which have a high incidence of pollution and density of tribal and forest population.

The *Vahinis* also play a watch-dog role by reporting instances of environmental pollution, deforestation, poaching, etc. They function under the charge of District Collectors, with the active cooperation of the State/Union Territory governments. This scheme is entirely financed by the Ministry of Environment and Forests. Seven Centres of Excellence have been set up by the Ministry to strengthen awareness, research and training in priority areas of Environmental Science and Management. These are: Centre for Ecological Sciences, Bangalore; Centre for Mining Environment, Dhanbad; Centre for Environmental Education, Ahmedabad; CPR Environmental Education Centre, Chennai; Salim Ali Centre for Ornithology and Natural History, Coimbatore; the Centre for Environmental Management of Degraded Ecosystems, Delhi, and the Tropical Botanical Garden and Research Institute, Thiruvananthapuram, Kerala. The National Museum of Natural History (NMNH) set up in New Delhi in 1978, is concerned with the promotion of non-formal education in the area of environment and conservation.

Besides permanent exhibit galleries on various aspects of environment, the museum also conducts temporary exhibitions and a large number of educational programmes and activities for school children, college youth and the general public. Three Regional Museums of Natural History have been established at Mysore, Bhopal and Bhubaneswar. The Indian Council for Forestry Research and Education is the focal point for forestry education and extension development in the country.

The Indira Gandhi National Forest Academy, Dehra Dun, imparts in-service professional training to Indian Forest Service (IFS) professionals. State forest service colleges provide training to the officers of the State Forest Service (SFS). The Indian Plywood Industries Research and Training Institute, Bangalore, organises short-term courses in the area of wood science. The Indian Institute of Forest Management, Bhopal, also provides training in forest management and allied subjects to persons from the Indian Forest Service, forest development corporations, and forest-related industries to develop forestry programmes. The Wildlife Institute of India, Dehradun, provides in- service training to forest officers, wildlife ecologists and other professionals for conservation and management of the wildlife resources of the country.

ENVIRONMENTAL INFORMATION SYSTEM

An Environmental Information System (ENVIS) was set up by the Ministry of Environment and Forests in 1982 as a decentralised information network for collection, storage, retrieval and dissemination of environmental information. Besides the Focal Point in the Ministry, ENVIS network presently consists of 25 subject-oriented centres known as ENVIS Centres, set up in various institutions/organisations of the country in priority areas of environment such as environmentally sound and appropriate technology, bio- degradation of wastes, desertification, estuary, mangroves, coral and lagoons, media and environment, environmental education, solid wastes disposal, animal ecology, Himalayan ecology, etc. The Focal Point brings out a quarterly abstracting journal, *Paryavaran Abstracts*, containing information about environmental research in the Indian context. ENVIS has been designated as National Focal Point (NFP) and Regional Service Centre (RSC) for South Asia sub-regional countries by INFOTERRA (an international referral system for sources of information on environment) of United Nations Environment Programmes (UNEP). The ENVIS network responded to a total of 12,756 queries during 1998-99 about a variety of environmental concerns.

FELLOWSHIPS AND AWARDS

- Instituted in 1987, the *Indira Gandhi Paryavaran Puraskar*, consisting of a cash component of rupees one lakh, a silver trophy and a citation, is awarded every year to an organisation and to an individual for significant contributions in the field of environment.
- The *Indira Priyadarshini Vrikshamitra Awards* were constituted by the Ministry of Environment and Forests in 1986 to recognise outstanding contributions of individuals and organisations in the field of afforestation and wastelands development. Starting

in 1993, 12 annual awards are given to individuals, educational institutions, panchayats, voluntary agencies, government agencies and the corporate sectors. Each award carries a medallion, a citation and a cash component of `50,000.

- The *Mahavriksha Puraskar*, instituted by the National Afforestation and Eco-development Board (NAEB) during 1993-94 is given every year to individuals/organisations for trees of notified species having the largest girth and height and in good health and vigour. The award consists of a cash prize of `25,000, a plaque and a citation.
- *Rajiv Gandhi Environment Award for Clean Technology* instituted in 1993, is given to industrial units that have made significant and measurable contributions towards development of clean technologies and innovative solutions to environmental problems created by industrial operations. With the objective of encouraging original and applied research among Group 'A' scientists in the Ministry of Environment and Forests and its associated offices and autonomous bodies, *Paryavaran Evam Van Mantralaya Vishisht Vaigyanik Puraskar* was instituted in 1992-93. The scheme provides every year for two awards worth `20,000 each.
- The *Pitambar Pant National Environment Fellowship Award* instituted in 1978, is awarded annually to recognise, encourage and support excellence in any branch of research related to environmental science. In recognition of the *Bishnoi* community's contribution to nature conservation and to encourage studies on desert ecology, the Ministry has instituted a Desert Ecology Fellowship at the University of Jodhpur.
- Instituted in 1995, the *B. P. Pal National Environment Fellowship Award for Biodiversity* is in recognition of significantly important research and development contributions and is intended to encourage talented individuals to devote themselves whole time to R&D pursuits in the field of bio-diversity.
- The *Rajiv Gandhi National Wildlife Conservation Award* and Salim Ali/ Kailash Sankhla Fellowships were instituted recently by the Ministry for recognising eminent officers and field workers for exemplary work in the field of Wildlife Conservation and Research. Award in the name of Amrita Devi Vishnoi has been instituted for showing valour and courage in protection of wildlife.

ENVIRONMENTAL LEGISLATION

Major legislations directly dealing with the protection of environment are: the Wildlife (Protection) Act, 1972, the Forest (Conservation) Act, 1980, the Water (Prevention and Control of Pollution) Act, 1974, the Water (Cess) Act, 1977, the National Environment Appellate Authority Act, 1977, the Air (Prevention and Control of Pollution) Act, 1981, the Environment (Protection) Act, 1986, the Public Liability Insurance Act, 1991 and the National Environment Tribunal Act, 1995. The Wildlife (Protection) Act, 1972 provides for rational and modern wildlife management while the Forest (Conservation) Act, 1980 has been enacted to check indiscriminate deforestation/diversion of forest land for non-forest purposes. The Water and Air Acts are the major instruments for the control of water and air pollution and these have provided for the establishment of the Central and State Pollution Control Boards.

The Environment (Protection) Act, 1986 is a landmark legislation which provides for a single focus in the country for the protection of environment and aims at plugging the loopholes in the existing legislation. The Public Liability Insurance Act, 1991, provides for mandatory insurance for the purpose of providing immediate relief to the persons affected by accidents occurring while handling any hazardous substances. The National Environment Tribunal Act, 1995, seeks to constitute a Tribunal with Benches to award compensation for damage to persons, property and the environment arising out of any activity involving hazardous substances. The National Environment Appellate Authority Act, 1977 provides for the establishment of a National Environment Appellate Authority to hear appeals against environmental clearance given by the Ministry.

All these Acts are amended from time to time to rationalise and expand their scope, coverage and penal provisions. The Ministry is preparing a comprehensive National Policy on Environment. The Water (Prevention and Control of Pollution) Cess Act, 1977 is also being amended.

INTERNATIONAL COOPERATION

The Ministry of Environment and Forests functions as a nodal agency for United Nations Environment Programmes (UNEP), South Asia Cooperation Environment Programmes (SACEP), and International Centre for Integrated Mountain and Development (ICIMOD), International Union for Conservation of Nature and Natural Resources (IUCN) and various international agencies, regional bodies and multilateral institutions.

India is signatory to the following important international treaties/agreements in the field of environment:

- International Convention for the regulation of Whaling
- International Plant Protection Convention
- The Antarctic Treaty
- Convention on Wetlands of international importance
- Convention on International trade in Endangered Species of Wild Flora and Fauna
- Protocol of 1978 relating to the international convention for the prevention of pollution from ships
- Vienna Convention for the protection of the Ozone Layer
- Convention on Migratory Species
- Basel Convention on Trans-boundary movement of hazardous substances
- Framework Convention on Climate Change
- Convention on conservation of bio-diversity
- Montreal Protocol on the substances that deplete the ozone layer and International Convention for Combating Desertification

The Ministry and its agencies receive assistance from various countries such as Sweden, Netherlands, Norway, Denmark, Australia, U.K., U.S.A, Canada, Japan, FRG, etc. , on bilateral basis and from several UN and other multilateral agencies such as UNDP, World Bank, Asian Development Bank OECF (Japan) and ODA (U.K.) for various environmental and forestry projects

OZONE CELL

Global efforts to protect the ozone layer started in the early seventies leading to the adoption of the Vienna Convention on Ozone Depleting Substances (ODS) in 1985 and the Montreal Protocol in 1987. India acceded to the Montreal Protocol, along with its London Amendment in 1992. To meet the country's commitment on ODS phase-out under the protocol and to disseminate information on ozone and ODS, the Ministry has established an Ozone Cell.

A newsletter on ozone issues is being published by the Cell every two months and a number of workshops and seminars are also conducted to create awareness about ozone among industries. UNEP-IE Ozone Action Programmes, Paris held its first south-Asia ODS officers Network Meeting in Delhi. Swiss Development Corporation organised a strategy workshop on service skill development in ecological refrigeration. The Ministry provides custom/excise duty exemption for ODS phase-out projects and detailed guidelines/procedures have been finalised to grant duty exemption for new investments with non-ODS technologies. A policy to issue licences for import of ODS has been implemented and the Reserve Bank of India, on the recommendation of the Ministry, has issued instructions to all commercial banks prohibiting finance or refinance of new investments with ODS technologies. One hundred seventy seven projects worth about US $ 49 million have been approved for India by the Multilateral Fund under the convention. Trade in ODS with non-Parties has been banned. Imports and exports of ODS have been licensed. Exports of CFCs to developed countries has been stopped. Draft rules on Ozone Depleting Substances (ODS) phaseout were published on 9 May 1998 in the Gazette of India. Policy to issue licences for imports of ODS has been implemented.

CHRONOLOGICAL HIGHLIGHTS

- 1951 National Forest Policy of 1894 replaced *Van Mahotsava* (national tree planting festival launched)
- 1952 Indian Wildlife Board Constituted
- 1953 Central Soil Conservation Board set up
- 1962 Animal Welfare Board of India established
- 1965 Central Forestry Commission set up to study the implementation of the National Forest Policy by State Governments and Union Territories
- 1966 Indian Forest Service constituted
- 1973 Project Tiger initiated
- 1980 Separate Department of Environment
- 1981 Forest Survey of India (FSI) set up
- 1983 National Wildlife Action Plan adopted
- 1985 Department of Environment and Forest created Central Ganga
- Authority set up
- 1987 Indira Gandhi Paryavaran Puraskar instituted

❐

7

Chemical Processes for Environmental Engineering

STUDY

Environmental chemistry is the study of chemical processes occurring in the environment which are impacted by humankind's activities. These impacts may be felt on a local scale, through the presence of urban air pollutants or toxic substances arising from a chemical waste site, or on a global scale, through depletion of stratospheric ozone or global warming. The focus in our courses and research activities is upon developing a fundamental understanding of the nature of these chemical processes, so that humankind's activities can be accurately evaluated.

The field of environmental chemistry is both very broad and highly interdisciplinary. Within the Department of Chemistry we have a core group of faculty whose research interests are in atmospheric and aquatic chemistry, photochemistry, and the chemistry and transport of long-lived pollutants. We interact with other chemists in the Department, with numerous other researchers at the University who have related interests, and with nearby government agencies. Indeed, the setting for the study of environmental chemistry is ideal. Our graduate Programmes consists of graduate courses which stress the fundamental photochemical, kinetic, analytical and transport aspects of environmental phenomena, regular seminars, and close interactions between the different research groups. We emphasize that students are able to put expertise in their own research field into a global context.

The field of environmental chemistry is rapidly expanding, and excellent employment opportunities exist in the academic, government, industrial and public policy sectors. The term environment refers to the natural elements that make up the Earth and surround living organisms, enabling life. The study of the environment and its proper conservation, has gained enormous importance in last three decades. The term environmental pollution has become one of the most important issues of global concern.

Accessible Components of Environment

- *Atmosphere*: This is a blanket of gaseous layer around the Earth.
- *Hydrosphere*: This includes all sources of water like oceans, rivers, lakes, glaciers, ground water, etc. and comprises about 96% of Earth's surface.
- *Lithosphere*: This comprises of soil, minerals, organic matter etc. and refers to Earth's solid crust containing the outer mineral cover.
- *Biosphere*: This is the domain of living organisms in contact with atmosphere, hydrosphere as well as lithosphere.

Man has begun to effect significant changes in his natural environment mainly after the advent of the Industrial revolution. The change in industrial and agricultural practices in the last 300 years and the direction of overall economic, technological and institutional development, has led to a tremendous increase in the use of energy and its impact on the environment. Due to these climatic changes, health hazards, effects on food and water supplies, changing habitats and eco-systems are effecting plants and animals life. Effects on the environment such as ozone depletion, greenhouse effect, acid rains, etc. have surfaced in recent years and are often the hot topics of discussion. The branch of chemistry, which deals with the study of various chemical phenomena occurring in the environment, is called environmental chemistry.

CONCEPT AND SCOPE

CONCEPT

Environmental chemistry" is the scientific study of the chemical and biochemical phenomena that occur in natural places. It should not be confused with green chemistry, which seeks to reduce potential pollution at its source. It can be defined as the study of the sources, reactions, transport, effects, and fates of chemical species in the air, soil, and water environments; and the effect of human activity on these. Environmental chemistry is an interdisciplinary science that includes atmospheric, aquatic and soil chemistry, as well as heavily relying on analytical chemistry and being related to environmental and other areas of science.

Environmental chemistry is that branch of one,which deals with the study of chemical and biochemical phenomena that occur in natural places like air,soil,water.

Environmental chemistry involves first understanding how the uncontaminated environment works, which chemicals in what concentrations are present naturally, and with what effects. Without this it would be impossible to accurately study the effects humans have on the environment through the release of chemicals.

Environmental chemists draw on a range of concepts from chemistry and various environmental sciences to assist in their study of what is happening to a chemical species in the environment. Important general concepts from chemistry include understanding chemical reactions and equations, solutions, units, sampling, and analytical techniques.

SCOPE

In chemistry the term mole represents a pile or mass of atoms, molecules, ions or electrons. Just as common man measures quantity in terms of kilograms or dozens, a chemical scientist deals with a 'mole' of atoms, molecules, ions or electrons. A mole is the amount of a substance that contains the same number of units of atoms or molecules or ions, as there are atoms in 12 g of the carbon 12 C

The 12 g of carbon contains a certain number of atoms of carbon. This number is estimated to be 6.023 × 1023 and is called Avogadro's number. Thus 12 g of the carbon 12 C is one mole of carbon.

Since the number of atoms in 12 g of carbon is called Avogadro's number, a mole may be redefined as the amount of a substance, which contains Avogadro number of molecules.

Avogadro's number is the number of elementary units (atoms, molecules, or ions) present in one mole of a substance, and it is equal to 6.023×10^{23}.

The concept of a mole being equal to a number 6.023×10^{23} is akin to the concept of a dozen being equal to a number 12.

A mole of atoms consists of Avogadro's number of atoms, and has a mass equal to one gram atomic mass of that element.

Therefore, the following illustrations hold true:

- One mole of oxygen atoms contains 6.023×10^{23} atoms of oxygen and has a mass of 16 g.
- One mole of sodium atoms contains 6.023×10^{23} atoms of sodium, and has a mass of 23g.

Similarly one mole of molecule of a substance has Avogadro's number, $\times 10^{23}$, of molecules and has a mass equal to one gram molecular mass of that substance.

- One mole of oxygen molecules contains 6.023×10^{23} molecules of oxygen and has a mass of 32 g.
- One mole of sodium (mono atomic) contains 6.032×10^{23} molecules (or atoms, as it is mono atomic) and has a mass of 23 g.
- One mole of sulphuric acid contains 6.023×10^{23} molecules of sulphuric acid and has a mass of 98 g.
- One mole of carbon dioxide contains 6.023×10^{23} molecules of carbon dioxide, and has a mass of 44 g.
- One mole of an ionic substance like sodium chloride or calcium chloride contains 6.023×10^{23}formula units.
- One mole of sodium chloride contains 6.023×10^{23} formula units *i.e.*, 6.023×10^{23} sodium ions and 6.023×10^{23} chloride ions. It has a mass of 58.5 g (*i.e.*, 23 + 35.5) g.
- One mole of calcium chloride contains 6.023×10^{23} formula units *i.e.*, 6.023×10^{23} calcium ions and $[2 (6.023 \times 10^{23})]$ chloride ions. It has a mass of 111 g.
- One mole of aluminium chloride contains 6.023×10^{23} formula units of it, and contains 6.023×10^{23} aluminium ions and $(3 \times 6.023 \times 10^{23})$ chloride ions. It has a mass of 133.5 $(27 + \{3 \times 35.5\})$.

Normally, mass of 1 mole of a substance is called its molar mass. It is expressed as g/mol. For example molar mass of water is equal to 18 g/mol.)

COMPONENTS OF ENVIRONMENT

Components of environment can be Analysed at various levels *i.e.* At the level of activities, at the level of processes, and at the level of orientations. Some of these issues and examples are discussed below.

As discussed the following activities have been suggested as components of environmental quality:

- Fire protection.
- Comfort of home.
- Electric service.
- Privacy in your home.
- Relation with fellow workers.
- Postal service.
- Garbage-collection.
- Mechanical helpers in your family.
- Telephone services.
- Public-water system.

- Relation with Neighbours.
- Gas service.
- Freedom to live where you want.
- Sewage disposal service.
- Availability of food around your living place.
- Convenience for getting to important places.
- Noise level in the home setting.
- Beauty of your home.
- Security of your home.
- Topography of land around.
- Product available to the community.
- Medical care in your locality.
- Police protection.
- Quality of water used by the household.
- Natural outdoor recreation.
- Variety of wild life in community.
- Cleanliness of air around.
- Overall weather.
- Public information media.
- Crowding in your residential Neighbourhood.
- Transportation over long distance.
- Level of traffic congestion.
- Job-opportunities.
- Quality of water for recreation.
- Product quality and variety.
- Relations among group in the community.
- Freedom to move from class to class.
- Freedom to move form on job to another.
- Public services-gas, sewage etc.
- Man made environment.
- Physical condition of environment at school.
- Isolation of your community.
- Physical condition of environment where you work.
- Amount of open space around.
- Access to parks.
- Control of dogs and other pets.
- Cost of living.

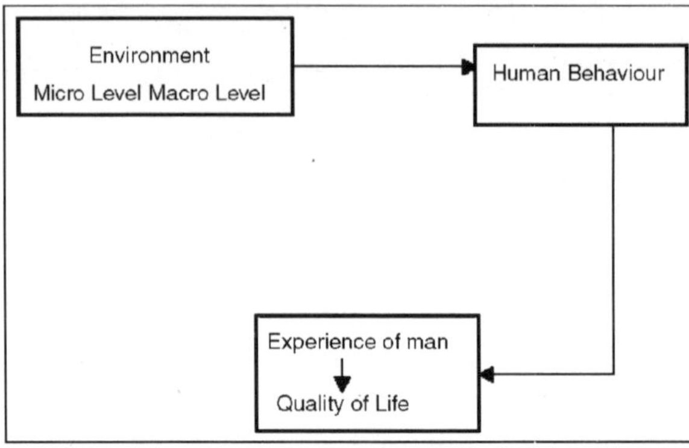

Understanding environment involves studying the complex relationships between the people and the typical physical settings in which they conduct their daily lives. Environmental interest in studying human behaviour in the familiar, everyday physical environment where people live and work as well as its relevance to the environment design and social planning has made it especially responsive to the demands of today's world.

Environmental studies are an area of social sciences where the focus of investigation is the interrelationship between the physical environment and human Behaviour and experience of man. Fig below presents this.

The other perspective of components of environment relates to environmental processes, its multidisciplinary and applied aspects.

ENVIRONMENTAL PROCESSES

Environment is new subject and a complex field of study. Therefore it is important to consider some of the characteristics that describe the what, how and why of the ways environmental processes work. In this chapter we attempt to explain the adaptation focus, physiological processes, the holistic view, interdisciplinary involvement and applied orientation of the subject matter.

Adaptation Focus

The process of adaptation. It is of interest to study how people adapt to the complex demands of the physical environment. For example, how people live in crowded places or overcrowded setting. Robert White defines adaptation as encompassing all the processes with their environment.

This includes most simple ways of dealing with minor environment and irritations to the most complex efforts to cope with major environment changes. These challenges in living system in interaction with the environment are the adaptation processes.

The holistic view of the organism and environment is considered along with the active role of living organism in relation to their environment.

PHYSIOLOGICAL PROCESSES

The adaptation focus of the environment emphasizes the process that mediates the effect of physical setting on human activity *e.g.* the effect of classroom noise on student's grades.

OVERALL VIEW

Historically, the analysis of the environment was at very small and molecular levels but now the emphasis has shifted from micro to macro level. It is how environment and Behaviour must be seen as interrelated parts of an indivisible whole.

The positive and adaptive ways in which people cope with environmental challenges suggest active roles. This view looks into how people have varied and creative ways to cope with their environments.

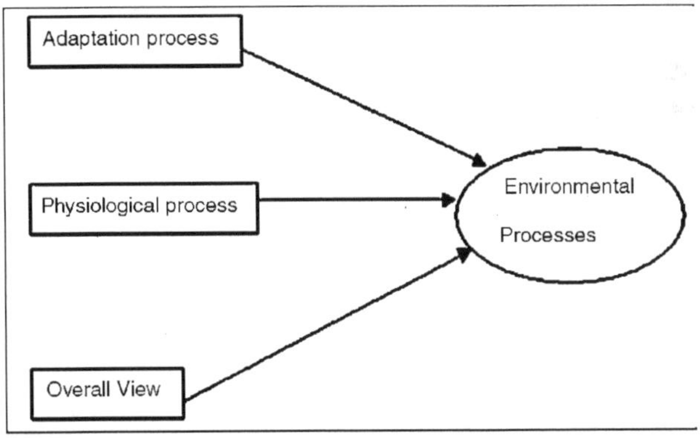

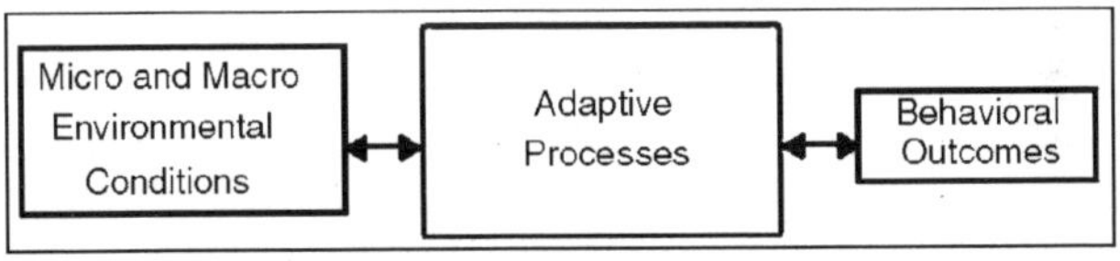

The adaptation process model (shown above) shows how environment effect on Behaviour is mediated by a number of processes.. It shows that the direction of effect in the environment-Behaviour relationship is reciprocal *i.e.* people may act on the environmental conditions while the environment in turn also acts on human Behaviour. The negative effects of situation such as over- crowding may be reversed through effective coping processes.

MULTIDISCIPLINARY ORIENTATION

The branches of study here have included not only environment but also other fields such as Sociology, Psychology, Geography, Anthropology, Medicine, Architecture and Planning etc. The study of human Behaviour in physical settings requires the works of researchers in many social sciences as well as of architects and planners responsible for the design of human settings. The areas of study include.

APPLIED ORIENTATION

The environmental study has orientation towards both the resolution of practical problem and the formulation of new theories. Lewins action research is a useful model for the intermingling of practical and theoretical needs of the environment.

VALUATION TECHNIQUES

Various techniques, which attempt to means environmental effects, are below.

Preventive Expenditure Techniques

How much people are prepared to spend to prevent degradation of environment can be used as a measure of the value of the environmental effects. It is, however, usually thought that this technique underestimates the value of the environmental effects for various reasons.

If the full scopes of environmental effect including the long-team consequences are beyond the common people's comprehension, the expenditure they would undertake to prevent degradation of environment would all very much short of the required levels. Secondly, it may always not be possible to fully offset all environmental effects by preventive expenditure especially in case of massive investment projects (such as large dams) which cause large environmental changes.

Replacement Cost Techniques

In this technique, how much people are prepared to spend to restore the environment to its previous state after degradation has occurred is taken as a measure of the value of the environmental effects. This technique also suffers from shortcomings similar to those mentioned in the context of the preventive expenditure technique. It is normally beyond the competence of common people to judge the full scope of damages done to the environment and it is not usually possible to restore the environment to its previous state after it has been damaged.

Property Value Techniques

Real estate values depend on various factors, including environmental factors. A residential accommodation at the lake or with a view to maintain commands a premium in the market. Air pollution depresses price of property in the area. People are willing to pay more for property similar in other respects in areas with clean air.

The difference in market price can be used as a measure of the value to society of the cleaner environment. The measurement of this value in practice may however be quite difficult. In order to separate the effect of cleaner environment on properly value from other factors (such as location, quality of neighbourhood, etc) that affect a site's value, one would require to collect data as the value of various sites throughout the area and analyse the data with the help of an appropriates statistical method.

There have been several studies exploring the relationship between air pollution and property values, most of these studies show a significant negative relationship between properly value and air pollution. These studies have, however, been criticized for underestimating the environmental effects of air pollution. People often do not know or understand the severity of pollution problems at different sites.

It requires special expertise to accurately estimate the pollution level at a site. Even if people are told about the presence of air pollutants in the area, they cannot make any use of such information because they lack knowledge about the effects of pollutants on human health or property. In such conditions it is not logical to expect that difference in properly value would correctly estimate the value of environmental effects of air pollution.

Human Health - Effect Technique

One method to estimate the value of environmental effects as to study the effect of pollutant exposures on human health.

It involves putting a monetary value on illness and death caused by exposure to pollutants. This idea of putting money value on human life does not appeal to everyone. Many regard human life to be of infinite value. But, is society's behaviour always consistent with such thinking? IS society prepared to pay whatever is necessary to prevent a death? If human life were considered of infinite value, we would never allow those activities which pose a risk to human beings, given that virtually no economic activity is completely risk- free, and they all would be stopped.

For example, injurious and deaths caused by road accidents could be reduced significantly if laws pertaining to road safely are made very strict. Several studies have shown that appropriate changes in design of vehicles and lowering of speed limits can reduce the number and severity of traffic accidents significantly. But we know that society does not take all these required steps. This ofcourse, does not mean that human life is cheap. Most people consider it invaluable, if not of infinite value.

One way to determine the effects of exposure to pollutants on human health is to study the incidence of diseases among people. Of course, air pollution is not the role cause of diseases. There are many other explanatory variables such as age, sex, hygienic conditions, climate, living standard and dietary habits. In order to separate the effect of air pollution from other variables, a large body of data on the incidence of various diseases among different sections of the population in various places needs to be collected and analysed with the help of appropriate statistical methods. An alternative method of determining the effects of exposure of pollutants is to conduct Labouratory experiments on plants or animals. Plants animals are exposed to different levels and different kinds of pollutants to study their effects. Rats are often used as study objects for various reasons. The results of such studies, however, cannot be generalized too much. These studies cannot tell much about the effects of pollutants on human health as the human physiology is much different from that of rat or any other animal.

Once the relationship between pollutant exposures and human health effects is determined, the next step is to estimate the value of human health effects in monetary terms. The human health effect technique calculates the value of the health effects by looking at lost economic production and increased health-care costs. The opportunity costs of an individual's premature death can be calculated by estimating the discounted present value of his future earning if he had not died prematurely. This provides a measure of lost economic production and thus a measure of loss to society. In case of illness, the cost of care is added to the losses of earnings. This technique however does not value the psychic costs of ill health.

Many people find such procedure of calculating the worth of human life highly unsatisfactorily because according to this criterion a large section of people would command a very low economic value, especially the old people, disabled and those unemployed or in poorly paid jobs.

Travel Cost Method

This method estimates the time and money that people are willing to spend on 'eco-trips', *i.e.*, visits to places of environmental value such as mountains, forests, lakes and parks. This method, however, severely underestimates the worth of the environmental assets and services because of its highly narrow focus. It only looks at the entertainment, recreational aspects of environmental assets.

Contingent Valuation

In this method, the value of environmental assets and services is assessed through public opinion surveys. People are directly asked their valuation of the environment. They are asked how much thing is willing to pay to conserve the environments.

There are problems, however, with approach. People often lack knowledge in that area; therefore their responses do not accurately reflect their concerns. Secondly, there is a general inherent problem with Survey method *i.e.* the answer to a question very much depends on the way it is asked. If people realise that their answer will have effect on the decisions they may reply in strategic manner.

For example, if how much they are willing to pay is a hypothetical question (*i.e.*, no payments are involved), people will answer it in one way. On the other hand, if there is a probability that they may actually have to pay they will answer it in another way. In the later case, there would be a systematic underestimation of the value of the environment. If the understand that they do not have to pay, the polluter will pay, their responses are likely to overestimate the value of the environment.

If they are asked how much compensation they should be given for the environmental damages, they would certainly exaggerate the worth of the environment. For these reasons, it is quite difficult to interpret the survey responses and come up with a figure which genuinely represents people's valuation of the worth of the environment. Investigators should try to frame questions in such a manner which induces individuals to respond not as interested parties but as disinterested observers, as people concerned with the public interest.

For example The study of university housing environment is one research area where theoretical and practical objectives have been successfully combined. Thus one may understand the components of environment at various levels *i.e.* At the level of activities, at the level of processes and at the level of orientations.

ENVIRONMENT AND ECOLOGY

A Systems Perspective

In the earlier sections it has been explained that there are many components of the concept and these concerns have always been there ever since the pre-historic time. The fact is that environment and ecology do not stand in isolation. Man. Society and culture are affected and in turn affect these concerns.

These could be viewed as the systems analysis. A systems thinking suggest that all subsystems of the main system interact with each other and if one sub system doesn't make impact it affects the total system. The systems analysis could further be studied as a closed system analysis and an open system analysis. Fig below presents a closed system analysis.

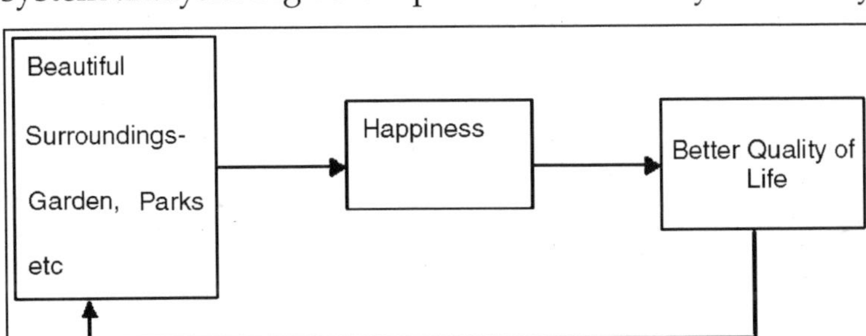

Similarly, there are ongoing impacts of environment on the quality of life of people, when time moves on. In fact, systems approach, views interactions and interdependence among subsystems are important as individual components. If one of the subsystems malfunctions, it affects the performance of the entire system.

Boulding (1956) has pointed out different levels of systems ranging from static structures to cybernetic structures and human systems. He believes that human systems and social systems are most complicated ones and we have yet to discover the rudiments of theoretical models for them.

Keeping these analogies and components analysis in mind one could further analyse ecology and environment within a systems perspective.

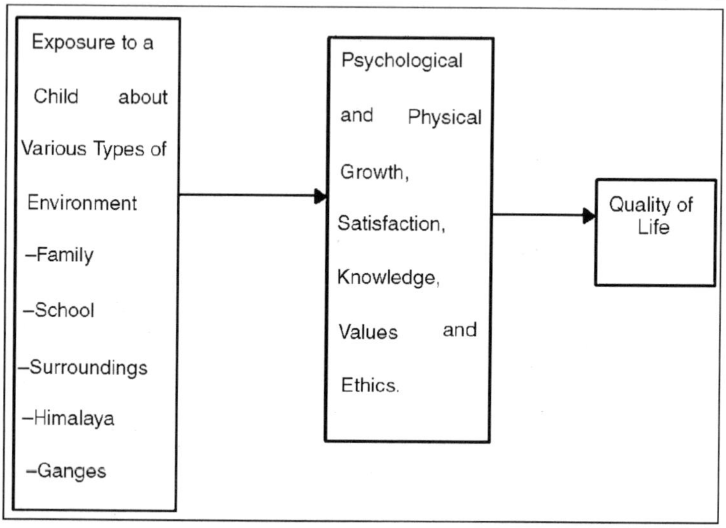

MULTIDISCIPLINARY PERSPECTIVE

Taking these as components of environment and ecology this web based course attempts to present a multidisciplinary system approach. The concepts are drawn from all subjects relevant to explain the environment scenario. Appropriate Technology, Psychology, Sociology, Economics, Management, and Philosophy, Values and Ethics are some subjects from where concepts and concerns are drawn. The course also presents methodological approaches, training, ideologies, globalization, ecological behaviour and knowledge management relating them to environment and ecology.

The 21st century is a century of shorter distances, closer places but one that is plagued by a multitude of problems, man made as well as natural disasters, catalyzed by man made actions.

The growing populace with its increased technology use is one that is seemingly empowered to control its actions and yet susceptible to the minutest happenings as were its forefathers.

Globalism - refers to an appreciation of the world as a single place or space and the ways in which people have internalized a basic view of the world as a common domain in which activities or events (natural or man-made)in one part might have profound effects in another, distant one. It is an awareness of large scale knock-on effects.

Today Ecology has moved from the realm of biology to become a stand alone discipline that encompasses all facets of human activity be they economic social or developmental. The word comes from the greek work for oikos, house and logi -logy. Ecology and environment encompass a number of disciplines. An overview of ecology suggests its multivariate nature.

Some of the ecological systems are:

- Community ecology
- Ecosystem ecology
- Biological ecology and diversity
- Geographical ecology
- Oceanographic and marine ecology
- Ecosystems of the world
- Evolutionary ecology
- Political ecology

ENVIRONMENT: A MULTIDISCIPLINARY PERSPECTIVE

Environment as a concept is this multidisciplinary involving a number of disciplines as presented earlier in fig which is repeated here.

Social Science	Technology
Economic	FactorsPolicy related factors
Social factors	Psychological factors

Environmental analysis also includes the understanding of multiple issues related to disciplines.

Some of the examples are:

- Air pollution
- Water pollution
- Industrial pollution
- Social environment
- Psychological environment
- Biological environment
- Ozone depletion
- Global climate change

- Rise in sea level and other natural changes
- Antarctica: a source for resource development
- Marine resources
- Offshore oil and gas
- Manmade habitats etc

This calls for:

- Studies of sustainable development
- Environmental politics
- Environmental audit
- Suitable environmental policies and legislation Therefore, in this section we have visualized an overall perspective of ecology and environment from multidisciplinary angle. The next section presents
- some of the methodological considerations to Analyse these issues.

ENVIRONMENT ANDECOLOGY ANALYSIS

Scientific Social Research

It would be difficult to comprehend the nature and content of research without an appreciation of the method we designate as scientific. Research, simply put is an endeavour to discover intellectual and practical answer to problems through the application of scientific methods to knowable universe. The scientific methods increase the likelihood of answering questions related to WHAT, HOW And Why in an enquiry. Einstein and Infield state the essence of science in the following words-"Science is the attempt of the human mind to find a connection between the world of ideas and world of phenomena. All the essential ideas in science were born in dramatic conflict between reality and our attempt at understanding the same." This does not mean, however, that science will always come out with an answer… Thompson observes "The belief that science has explained everything is hopeless misunderstanding." There may be many paths leading to truth, however scientific is entails more understandable and reliable. Meaning of Research therefore is 'A careful, critical enquiry or examination in seeking facts or principles in order to as certain something.

Scientific Social Research

It would be difficult to comprehend the nature and content of research without appreciation of the methods we designate as scientific. *According to Scientific Social Research have the following properties*: Reliance on empirical evidence Use of relevant concepts Commitment to objectivity Ethical neutrality Predictions based on probability Testing of conclusions through replication.

Research Process

There are various measures to study the effects of environment. Earlier strategies were not that sophisticated and reliable, thus mere have been fascinating development in this field.

- Co relational studies.
- Experimental studies in Labouratory settings.
- Small scale fields settings.
- Diverse research strategies.
- Evaluation studies.
- Econometric studies.
- Time series studies.
- *Correlation studies based census data*: It is based on study relationship between density and social pathology as reflected by cause data or over statistical records *e.g.* mere may be comparison between mortality, mental illness and serious crime in densely populated region and sparsely populated suburbs.

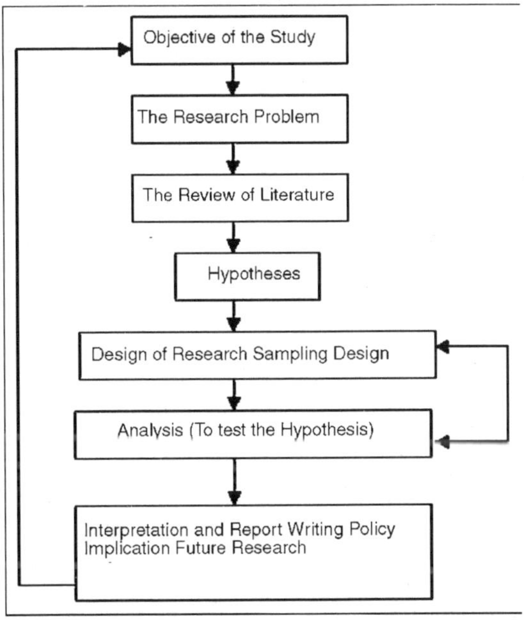

- *Experimental studies*: A more recent research strategy 1960's involves experiment in lab settings. Initial studies were dared on animals in controlled settings the god of this is study is to establish a casual like between crowding and social effects on humans. Later were also done as humans.
- *Study of environment in small-scale setting*: Most recent strategy such as - dormitories stores etc include - survey -questionnaires etc.

Diverse Research Strategies

Each strategy has its strength and weakness. In some animals studies it is the size of group that is more important man me space available to it also effect of density have inserted with other variables such as species type situation etc.

Approach for Environmental Studies

There are number of models and approaches to study environment and ecology. Measuring Environmental Values, impact assessment, people's reactions, environmental protection and crisis etc. there are many Economics based approaches issues and approaches.

They are:

- Market values
- Indirect methods
- Contingent valuation

:*Some of the indirect methods are*

- Preventive expenditure
- Replacement cost technique
- Travel cost method Other Methods include
- Technical coefficient and the value of Human Life
- Epidemiological studies
- Property value and Air pollutants
- Survey techniques

Environmental issues are too vital to be left out of economic decision making. Social cost benefit analysis cannot be complete without taking intconsideration environmental considerations. In this chapter we discuss variousmethods that have been suggested for measuring environmental values.

It is not easy to asses the environmental effects of a project. What types of physical environmental effects would a project cause and to what extent, can rarely be answered satisfactorily. There is always an element of uncertainty associated with there effects. Most large investment projects now require environmental impact assessment (EIAs) that estimates the environmental effects of the projects. Often these EIAs churn out estimates of the effects, which are not only incomplete but also, devoid of scientist rigour.

Even if the physical effects are assessed with reasonable degree of accuracy and certainty, the next problem one encounters is how to assign economic values to these effects. It is a widely held view that a market economy does not value all aspects of the environment appropriately. Markets and prices may not exist for many of the environmental effects that require valuation.

Even if market prices exist, they may be distorted for various reasons. The environmental effects of big projects are very large. In such cases, the project may actually alter prices. For these various reasons indirect methods are used for valuation of the environmental effects. There are a number of techniques developed by economists who attempt to value the environmental effects. Before discussing these various techniques it needs to be noted that total economic value (TEV) of environmental assets consists of use and no- use values. The total economic value has four components:

TEV = Direct use values + Indirect use values + Option value + Existence value

Let us consider the case of tropical forest. Direct values are the resources and services provided directly by the forest such as habitat and food for indigenous people, timber and minor non-timber products. Indirect use values of environmental assets are essentially their ecological functions, which indirectly support economic activity and human welfare. Tropical forests, for example, protect adjacent eco-systems, provide watershed protection, and are store of CO_2. An option value is the value placed on an option that allows for use of the environment in the future.

The amount individuals would be willing to pay to conserve a tropical forest, or at least some of its uses, for future would be its option value. Existence value of an environmental asset in the value placed on it independent of its any actual or possible future use. It relates to valuation of the resources as a unique asset in itself, an object of intrinsic value. Existence value of tropical forests arises because people simply want them to exist, even if they are never likely to ever see or use then in any way.

Environmental Impact Assessment: includes Environmental Inventory; Environmental Assessment; Environment Impact Statement; Historical Perspective; Environmental Assessment Process; Goals of EIA Process need to emphasize methodological issues related to following aspects:

- Air Environment.
- Water Environment.
- Noise Environment.
- Socio-Economic Environment.
- Cultural Environment.
- Biological Environment.
- People's Participation.
- Case Studies are also important in this regard. Some examples.
- Air Pollution Monitoring.

Ganga Action Plan

- *The study of Ganga*: The Holy River; Ganga; From Gomukh to Bay of Bengal; Ganga; Its importance, Ganga; Overuse, Abuse and Pollution; What is River Pollution? What are the Principal Sources of Pollution? How to measure the extent of pollution? Concern for the River; Formulation of the Ganga Action Plan; what does the Action Plan seek to do? River Front Facilities; Low Cost Sanitation; Construction of Electric Crematoria, Pollution from Industries; Scientific Content in the Context of Ganga Action Plan;Public Involvement in the Ganga Action Plan; Monitoring of Water Quality; Recycling Sewage; Where we went wrong?
- *Industrial Pollution*: Some Case Studies: Pollution Control Aspects of Fertilizer Industries; Ammonia Plant Effluents; Removal of Carbon; Oil Removal; Hydrogen Sulphide Removal; Nitrogen-Wash Tail Gas; Ammonium Sulphate Plant; Phosphoric Acid Plant; Solid Effluent; Liquid Effluent; Gaseous Effluent; Pollution Control in Petroleum Refineries and Petrochemical Units; Characteristics of Liquid Effluent; Refineries; Petrochemical Industries; Refinery Liquid Waste-Treatment Methods; Implant Measures; Steam Stripping; Oxidation; Removal of Phenols; Biological Treatment; Disposal of Sludge's; Treatment of Liquid Effluents from Petrochemical Industries.
- *Environmental Audit*: Case studies Indian Scenario Advantage of Audit; Environmental Audit. How to measure the extent of pollution? Formulation of the Ganga Action Plan; what does the Action Plan seek to do? River Front Facilities; Low Cost Sanitation; Construction of Electric Crematoria, Pollution from Industries; Scientific Content in the Context of Ganga Action Plan;Public Involvement in the Ganga Action Plan; Monitoring of Water Quality; Recycling Sewage; What do we need to do? Can people's participation help in this project?
- *Industrial Pollution*: Some Case Studies: Pollution Control Aspects of Fertilizer Industries; Ammonia Plant Effluents; Removal of Carbon; Oil Removal; Hydrogen Sulphide Removal; Nitrogen-Wash Tail Gas; Ammonium Sulphate Plant; Phosphoric Acid Plant; Solid Effluent; Liquid Effluent; Gaseous Effluent; Pollution Control in Petroleum Refineries and Petrochemical Units; Characteristics of Liquid Effluent; Refineries; Petrochemical Industries; Refinery Liquid Waste-Treatment Methods; Implant Measures; Steam Stripping; Oxidation; Removal of Phenols; Biological Treatment; Disposal of Sludge's; Treatment of Liquid Effluents from Petrochemical Industries.

CYCLES OF MATTER IN ENVIRONMENT

Environmental problems caused by human interference in matter cycles. In the past decades humans have been the cause of extensive environmental pollution. Pollution is not a one-sided term; all kinds of matter has been applied by humans, resulting in a disturbance in natural processes. Disturbances of natural processes are clearly shown, when you take a look at human interference in matter cycles. This has caused various environmental problems, which are important issues today.

HYDROLOGICAL CYCLE H_2O

There are three ways in which humans cause environmental problems by interfering with the hydrological cycle. Firstly, we remove large quantities of freshwater from rivers, lakes and groundwater supplies. In areas where great amounts of people use the water from groundwater supplies, groundwater may be fully depleted. This effect is enhanced by vegetation removal to create new farmland or underground mines, or to build roads. Secondly, because of vegetation removal, rainwater washes away and it no longer infiltrates, so that the groundwater supplies are not restored. In many cases groundwater supplies will than be filled up with salt water from rivers and lakes. When groundwater is salinated the entire area is influenced. Other effects also occur because of vegetation removal; the risk of flooding is enhanced, and soil erosion and land sliding increase. Finally, people change the quality of water, by adding nutrients and contaminants.

This causes the ecological processes that usually purify the water to be disturbed:

- *Precipitation*: Transported through the circulating atmosphere the clouds move themselves inland, as a result of gravity, and lose their water as it falls back unto the ground. This phenomenon is called rain or precipitation.
- *Infiltration*: Rainwater infiltrates into the ground and sinks to the saturated zone, where it becomes groundwater. Groundwater slowly moves from places of high elevation and pressure to places with low elevation and pressure. It moves from the area of infiltration through an aquifer and out to a discharge area, which can be either a sea or an ocean.
- *Transpiration*: Plants and other forms of vegetation take up water from the soil and excrete it again as water vapour. About 10% of the precipitation that falls on the ground vapourizes again through transpiration of plants, the rest evaporates from seas and oceans.
- *Surface run-off*: The rainwater that does not infiltrate into the soil will directly reach the surface water, as it will run-off to rivers

and lakes. After that it will be transported back to the seas and oceans. This water is called surface run-off.

- *Evaporation*: Due to the influence of sunlight the water in oceans and lakes will warm up. As a result of that it will evaporate and rise up into the atmosphere. There it will form clouds that will eventually cause rainwater to fall back on earth. The evaporation of oceans is the most important kind of evaporation.

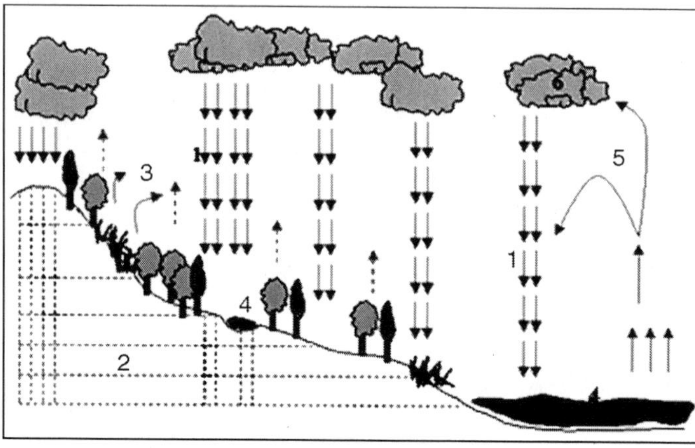

Fig. 7.1: The Hydrological Cycle

- *Condensation*: In contact with the atmosphere the water vapour will transform back to liquid, so that it will be visible in the air. These accumulations of water in the air are what we call clouds.

Carbon Cycle

The geological component of the carbon cycle is where it interacts with the rock cycle in the processes of weathering and dissolution, precipitation ofminerals, burial and subduction, and volcanism. In the atmosphere, carbonic acid forms by a reaction with atmospheric carbon dioxide (CO_2) and water. As this weakly acidic water reaches the earth as rain, it reacts with minerals at the earth's surface, slowly dissolving them into their component ions through the process of chemical weathering. These component ions are carried in surface waters like streams and rivers eventually to the ocean, where they precipitate out as minerals like calcite ($CaCO_3$). Through continued deposition and burial, this calcite sediment forms the rock called limestone. As seafloor carbon is pushed deeper into the earth by tectonic forces, it heats up, eventually melts, and can rise back up to the surface, where it is released as CO_2 and returned to the atmosphere. This return to the atmosphere can occur violently through volcanic eruptions, or more gradually in seeps, vents, and CO_2-rich hotsprings. Tectonic uplift can also expose previously buried limestone.

One example of this occurs in the Himalayas where some of the world's highest peaks are formed of material that was once at the bottom of the ocean. Weathering, subduction, and volcanism control atmospheric carbon dioxide concentrations over time periods of hundreds of millions of years.

Biological Carbon Cycle

Biology plays an important role in the movement of carbon between land, ocean, and atmosphere through the processes of photosynthesis and respiration. Virtually all multicellular life on Earth depends on the production of sugars from sunlight and carbon dioxide (photosynthesis) and the metabolic breakdown (respiration) of those sugars to produce the energy needed for movement, growth, and reproduction. Plants take in carbon dioxide (CO_2)

from the atmosphere during photosynthesis, and release CO_2 back into the atmosphere during respiration through the following chemical reactions:

Respiration: $C_6H_{12}O_6$ (organic matter) + $6O_2 \rightarrow 6CO_2 + 6\ H_2O$ + energy

Photosynthesis: Energy (sunlight) + $6CO_2 + H_2O \rightarrow C_6H_{12}O_6 + 6O_2$

Through photosynthesis, green plants use solar energy to turn atmospheric carbon dioxide into carbohydrates (sugars). Plants and animals use these carbohydrates (and other products derived from them) through a process called respiration, the reverse of photosynthesis. Respiration releases the energy contained in sugars for use in metabolism and changes carbohydrate "fuel" back into carbon dioxide, which is in turn released to back to the atmosphere. The amount of carbon taken up by photosynthesis and released back to the atmosphere by respiration each year is about 1,000 times greater than the amount of carbon that moves through the geological cycle on an annual basis.

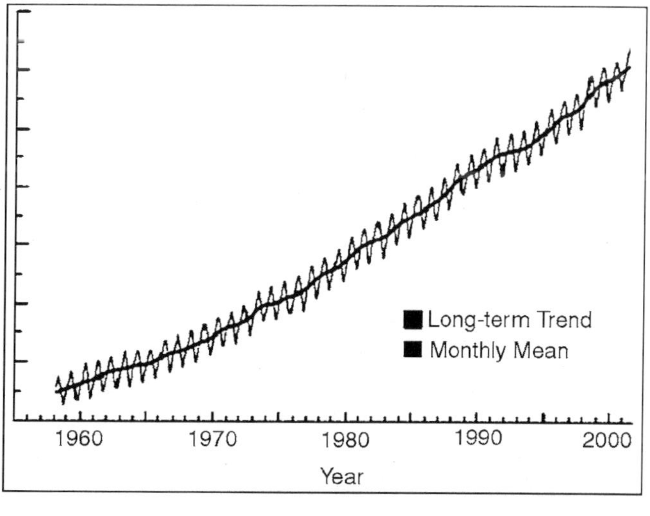

On land, the major exchange of carbon with the atmosphere results from photosynthesis and respiration. During daytime in the growing season, leaves absorb sunlight and take up carbon dioxide from the atmosphere. At the same time plants, animals, and soil microbes consume the carbon in organic matter and return carbon dioxide to the atmosphere. Photosynthesis stops at night when the sun cannot provide the driving energy for the reaction, though respiration continues. This kind of imbalance between these two processes is reflected in seasonal changes in the atmospheric CO_2 concentrations. During winter in the northern hemisphere, photosynthesis ceases when many plants lose their leaves, but respiration continues. This condition leads to an increase in atmospheric CO_2 concentrations during the northern hemisphere winter. With the onset of spring, however, photosynthesis resumes and atmospheric CO_2 concentrations are reduced. This cycle is reflected in the monthly means of atmospheric carbon dioxide concentrations shown in Figure.

In the oceans, phytoplankton (microscopic marine plants that form the base of the marine food chain) use carbon to make shells of calcium carbonate (CaCO3). The shells settle to the bottom of the ocean when phytoplankton die and are buried in the sediments.

The shells of phytoplankton and other creatures can become compressed over time as they are buried and are often eventually transformed into limestone. Additionally, under certain geological conditions, organic matter can be buried and over time form deposits of the carbon-containing fuels coal and oil. It is the non-calcium containing organic matter that is transformed into fossil fuel. Both limestone formation and fossil fuel formation are biologically controlled processes and represent long-term sinks for atmospheric CO_2.

ALTERATION OF CARBON CYCLE

Recently, scientists have studied both short- and long-term measurements of atmospheric CO_2 levels. Charles Keeling, an oceanographer at the Scripps Institute of Oceanography, is responsible for creating the longest continuous record of atmospheric CO_2 concentrations, taken at the Mauna Loa observatory in Hawaii. His data (now widely known as the "Keeling curve," shown in Figure) revealed that human activities are significantly altering the natural carbon cycle.

Since the onset of the industrial revolution about 150 years ago, human activities such as the burning of fossil fuels and deforestation have accelerated, and both have contributed to a long-term rise in atmospheric CO_2. Burning oil and coal releases carbon into the atmosphere far more rapidly than it is being removed, and this imbalance causes atmospheric carbon dioxide concentrations to increase.

In addition, by clearing forests, we reduce the ability of photosynthesis to remove CO_2 from the atmosphere, also resulting in a net increase. Because of these human activities, atmospheric carbon dioxide concentrations are higher today than they have been over the last half-million years or longer. Because CO_2 increases the atmosphere's ability to hold heat, it has been called a "greenhouse gas." Scientists believe that the increase in CO_2 is already causing important changes in the global climate.

Many attribute the observed 0.6 degree C increase in global average temperature over the past century mainly to increases in atmospheric CO_2. Without substantive changes in global patterns of fossil fuel consumption and deforestation, warming trends are likely to continue.

The best scientific estimate is that global mean temperature will increase between 1.4 and 5.8 degrees C over the next century as a result of increases in atmospheric CO_2 and other greenhouse gases.

This kind of increase in global temperature would cause significant rise in average sea-level (0.09-0.88 meters), exposing low-lying coastal cities or cities located by tidal rivers such as New Orleans, Portland, Washington, and Philadelphia to increasingly frequent and severe floods. Glacial retreat and species range shifts are also likely to result from global warming, and it remains to be seen whether relatively immobile species such as trees can shift their ranges fast enough to keep pace with warming.

Even without the changes in climate, however, increased concentrations of CO_2could have an important impact on patterns of plant growth worldwide.

Because some species of plants respond more Favourably to increases in CO_2 than others, scientists believe we may see pronounced shifts in plant species as a result of increasing atmospheric CO_2 concentrations, even without any change in temperature.

For example, under elevated CO_2 conditions, shrubs are thought to respond more Favourably than certain grass species due to their slightly different photosynthetic pathway. Because of this competitive inequality, some scientists have hypothesized that grasslands will be invaded by CO_2-responsive grass species or shrubby species as CO_2 increases.

In an attempt to understand whether recently observed changes in the global carbon cycle are a new phenomenon, or have instead occurred throughout geologic history, scientists have devoted considerable effort to developing methods for understanding Earth's past atmosphere and climate.

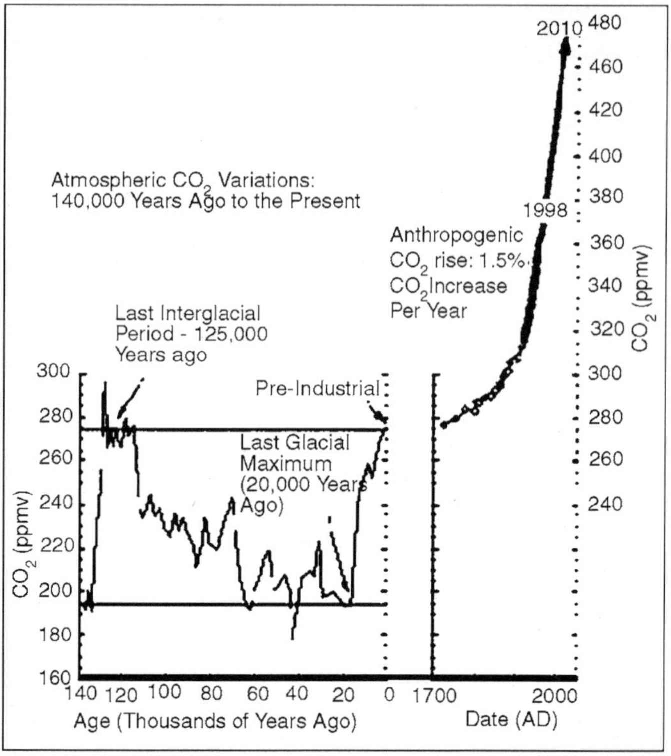

These techniques include analysis of gas bubbles trapped in ice, tree rings, and ocean and lake floor sediments for clues about past climates and atmospheres. Together, these techniques suggest that over the past 20 million years, the Earth's climate has oscillated between relatively warm and relatively cold conditions called interglacial and glacial periods. During interglacial periods, atmospheric CO_2 concentrations were relatively high, and during glacial periods, CO_2 concentrations were relatively low. We are currently in an interglacial warm period, and human activities are pushing CO_2 concentrations higher than they have been for hundreds of thousands of years (Figure).

Understanding and mitigating negative impacts of atmospheric CO_2 enrichment constitute two of the most central challenges that environmental scientists and policy makers currently face. In order to address this issue, the scientific community has formed the Intergovernmental Panel on Climate Change (IPCC), an international, interdisciplinary consortium comprised of thousands of climate experts colLabourating to produce consensus reports on climate change science.

Many nations have agreed to conditions specified by the Kyoto Protocol, a multilateral treaty aimed at averting the negative impacts associated with human-induced climate change. The country, which is currently responsible for approximately one quarter of global CO_2 emissions, has so far declined to participate in the Kyoto Protocol.

NITROGEN CYCLE

The nitrogen cycle involves the uptake of nitrogen form the atmosphere by a process called fixation which is carried out by microbes or industrial processes. Decomposition of biological waste by microbes can return nitrogen to the atmosphere. Nitrogen is mainly used by humans as a fertilizer in farmlands, but its excessive usage can lead to serious problems (such as eutrophication).

Nitrogen (N) is an essential component of DNA, RNA, and proteins, the building blocks of life. All organisms require nitrogen to live and grow. Although the majority of the air we breathe is N_2, most of the nitrogen in the atmosphere is unavailable for use by organisms. This is because the strong triple bond between the N atoms in N_2 molecules makes it relatively inert.

In fact, in order for plants and animals to be able to use nitrogen, N_2 gas must first be converted to more a chemically available form such as ammonium (NH_4^+), nitrate (NO_3^-), or organic nitrogen (*e.g.* urea–$(NH_3)_2CO$).

The inert nature of N_2 means that biologically available nitrogen is often in short supply in natural ecosystems, limiting plant growth and biomass accumulation. Nitrogen is an incredibly versatile element, existing in both inorganic and organic forms as well as many different oxidation states. The movement of nitrogen between the atmosphere, biosphere, and geosphere in different forms is described by the nitrogen cycle (Figure below), one of the major biogeochemical cycles.

Similar to the carbon cycle, the nitrogen cycle consists of various storage pools of nitrogen and processes by which the pools exchange nitrogen (arrows). Five main processes cycle nitrogen through the biosphere, atmosphere, and geosphere: nitrogen fixation, nitrogen uptake (organismal growth), nitrogen mineralization (decay), nitrification, and denitrification. Microorganisms, particularly bacteria, play major roles in all of the principal nitrogen transformations.

As microbially mediated processes, these nitrogen transformations tend to occur faster than geological processes like plate motion, a very slow, purely physical process that is a part of the carbon cycle.

Instead, rates are affected by environmental factors that influence microbial activity, such as temperature, moisture, and resource availability.

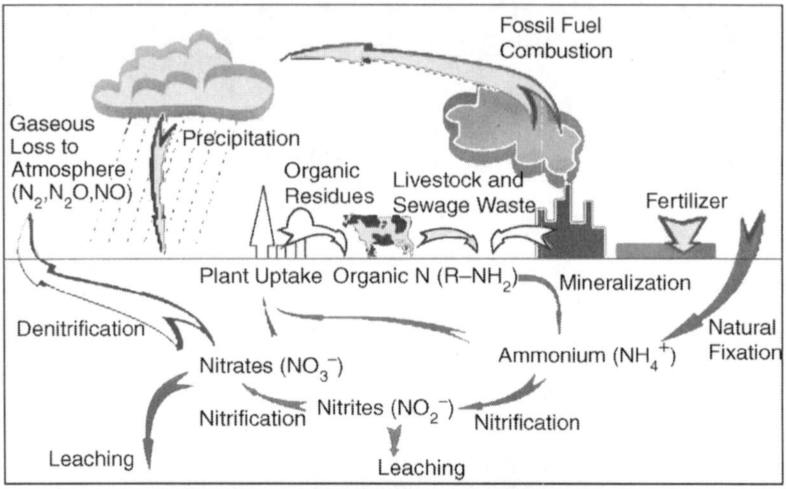

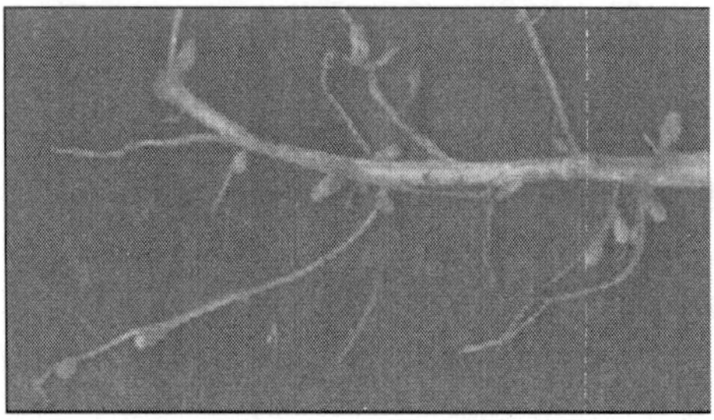

Nitrogen Fixation

$N_2 \rightarrow NH_4$+ Nitrogen fixation is the process wherein N_2 is converted to ammonium, essential because it is the only way that organisms can attain nitrogen directly from the atmosphere. Certain bacteria, for example those among the genus Rhizobium, are the only organisms that fix nitrogen through metabolic processes.

Nitrogen fixing bacteria often form symbiotic relationships with host plants. This symbiosis is well-known to occur in the legume family of plants beans, peas, and clover). In this relationship, nitrogen fixing bacteria inhabit legume root nodules (Figure) and receive carbohydrates and a Favourable environment from their host plant in exchange for some of the nitrogen they fix.

There are also nitrogen fixing bacteria that exist without plant hosts, known as free-living nitrogen fixers. In aquatic environments, blue-green algae (really a bacteria called cyanobacteria) is an important free-living nitrogen fixer.

In addition to nitrogen fixing bacteria, high-energy natural events such as lightning, forest fires, and even hot lava flows can cause the fixation of smaller, but significant amounts of nitrogen (Figure above).

The high energy of these natural phenomena can break the triple bonds of N_2 molecules, thereby making individual N atoms available for chemical transformation.

Within the last century, humans have become as important a source of fixed nitrogen as all natural sources combined. Burning fossil fuels, using synthetic nitrogen fertilizers, and cultivation of legumes all fix nitrogen. Through these activities, humans have more than doubled the amount of fixed nitrogen is pumped into the biosphere every year (Figure below), the consequences of which are discussed below.

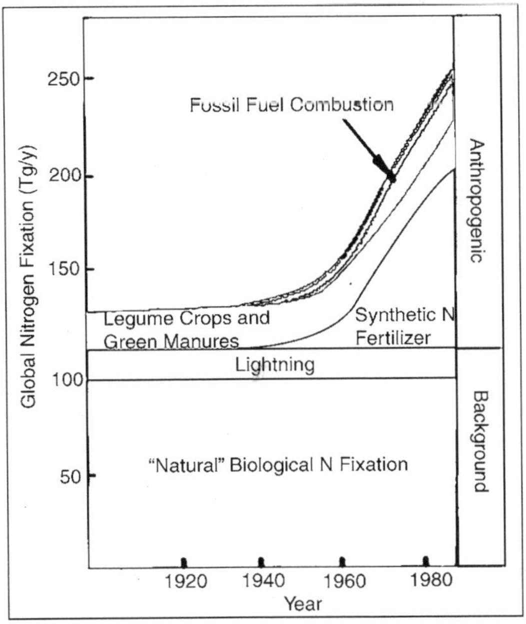

Nitrogen Uptake

NH_4+ → Organic N The ammonia produced by nitrogen fixing bacteria is usually quickly incorporated into protein and other organic nitrogen compounds, either by a host plant, the bacteria itself, or another soil organism. When organisms nearer the top of the food chain (like us!) eat, we are using nitrogen that has been fixed initially by nitrogen fixing bacteria.

Nitrogen Mineralization

Organic N → NH_4^+ After nitrogen is incorporated into organic matter, it is often converted back into inorganic nitrogen by a process called nitrogen mineralization, otherwise known as decay. When organisms die, decomposers (such as bacteria and fungi) consume the organic matter and lead to the process of decomposition. During this process, a significant amount of the nitrogen contained within the dead organism is converted to ammonium. Once in the form of ammonium, nitrogen is available for use by plants or for further transformation into nitrate (NO_3^-) through the process called nitrification.

Nitrification

$$NH_4^+ \rightarrow NO_3^-$$

Some of the ammonium produced by decomposition is converted to nitrate via a process called nitrification. The bacteria that carry out this reaction gain energy from it. Nitrification requires the presence of oxygen, so nitrification can happen only in oxygen-rich environments like circulating or flowing waters and the very surface layers of soils and sediments.

The process of nitrification has some important consequences. Ammonium ions are positively charged and therefore stick (are sorbed) to negatively charged clay particles and soil organic matter. The positive charge prevents ammonium nitrogen from being washed out of the soil (or leached) by rainfall. In contrast, the negatively charged nitrate ion is not held by soil particles and so can be washed down the soil profile, leading to decreased soil fertility and nitrate enrichment of downstream surface and groundwater's.

Denitrification

$NO_3^- \rightarrow N_2 + N_2O$ Through denitrification, oxidized forms of nitrogen such as nitrate and nitrite (NO_2^-) are converted to dinitrogen (N_2) and, to a lesser extent, nitrous oxide gas. Denitrification is an anaerobic process that is carried out by denitrifying bacteria, which convert nitrate to dinitrogen in the following sequence:

$$NO_3^- \rightarrow NO_2^- \rightarrow NO \rightarrow N_2O \rightarrow N_2.$$

Nitric oxide and nitrous oxide are both environmentally important gases. Nitric oxide (NO) contributes to smog, and nitrous oxide (N_2O) is an importantgreenhouse gas, thereby contributing to global climate change.

Once converted to dinitrogen, nitrogen is unlikely to be reconverted to a biologically available form because it is a gas and is rapidly lost to the atmosphere.

Denitrification is the only nitrogen transformation that removes nitrogen from ecosystems (essentially irreversibly), and it roughly balances the amount of nitrogen fixed by the nitrogen fixers described above.

ALTERATION OF N CYCLE AND ITS CONSEQUENCES

Early in the 20th century, a German scientist named Fritz Haber figured out how to short circuit the nitrogen cycle by fixing nitrogen chemically at high temperatures and pressures, creating fertilizers that could be added directly to soil. This technology has spread rapidly over the past century, and, along with the advent of new crop varieties, the use of synthetic nitrogen fertilizers has led to an enormous boom in agricultural productivity.

For example, on nitrogen-poor serpentine soils of northern Californian grasslands, plant assemblages have historically been limited to native species that can survive without a lot of nitrogen. There is now some evidence that elevated levels of atmospheric N input from nearby industrial and agricultural development have paved the way for invasion by non-native plants. As noted earlier, NO is also a major factor in the formation of smog, which is known to cause respiratory illnesses like asthma in both children and adults.

Currently, much research is devoted to understanding the effects of nitrogen enrichment in the air, groundwater, and surface water. Scientists are also exploring alternative agricultural practices that will sustain high productivity while decreasing the negative impacts caused by fertilizer use. These studies not only help us quantify how humans have altered the natural world, but increase our understanding of the processes involved in the nitrogen cycle as a whole.

Nitrogen is a part of vital organic compounds in microrganisms, such as amino acids, proteins and DNA. The gaseous form of nitrogen (N_2), makes up 78% of the troposphere.

One might think this means we always have plenty of nitrogen available, but unfortunately it does not work that way. Nitrogen in the gaseous form cannot be absorbed and used as a nutrient by plants and animals; it must first be converted by nitrifying bacteria, so that it can enter food chains as a part of the nitrogen cycle. During the conversion of nitrogen cyano bacteria will first convert nitrogen into ammonia and ammonium, during the nitrogen fixation process. Plants can use ammonia as a nitrogen source.

Nitrogen fixation is carried out according to the following reaction:

$$N_2 + 3\,H_2 \rightarrow 2\,NH_3$$

After ammonium fixation, the ammonia and ammonium that is formed will be transferred further, during the nitrification process. Aerobic bacteria use oxygen to convert these compounds. Nitrosomonas bacteria first convert nitrogen gas to nitrite (NO_2^-) and subsequently nitrobacter convert nitrite to nitrate (NO_3^-), a plant nutrient.

Nitrification is carried out according to the following reactions:

$$2\,NH_3 + 3O_2 \rightarrow 2\,NO_2 + 2\,H^+ + 2\,H_2O$$

$$2\,NO_2^- + O_2 \rightarrow 2\,NO_3^-$$

Plants absorb ammonium and nitrate during the assimilation process, after which they are converted into nitrogen-containing organic molecules, such as amino acids and DNA. Animals cannot absorb nitrates directly. They receive their nutrient supplies by consuming plants or plant-consuming animals.

When nitrogen nutrients have served their purpose in plants and animals, specialized decomposing bacteria will start a process called ammonification, to convert them back into ammonia and water-soluble ammonium salts. After the nutrients are converted back into ammonia, anaerobic bacteria will convert them back into nitrogen gas, during a process called denitrification.

Denitrification is carried out according to the following reaction:

$$NO_3^- + CH_2O + H^+ \rightarrow ½\, N_2O + CO_2 + 1½\, H_2O$$

Finally, nitrogen is released into the atmosphere again. The whole process starts over after release. A schematic representation of the nitrogen cycle is shown here:

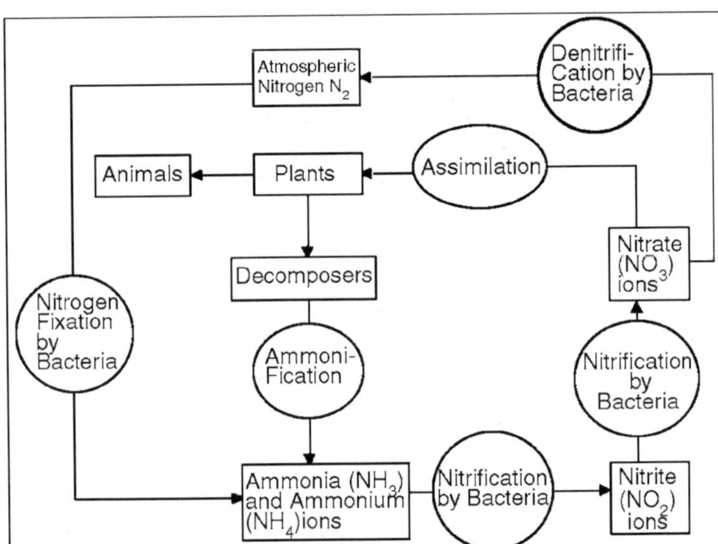

Nitrogen as a Limiting Factor

Although the nitrogen conversion processes often occurs and large quantities of plant nutrients are produced, nitrogen is often a limiting factor for plant growth. Water flowing across the soil causes this error. Nitrogen nutrients are water-soluble and as a result they are easily drained away, so that they are no longer available for plants.

Annamox Reaction

In 1999 researchers at the Gist-Brocades in Delft, The Netherlands, discovered a new reaction to be added to the nitrogen cycle; the so-called *annamox reaction*. This is now found to occur in the Black Sea, as well. The reaction implies conversion of nitrite and ammonium to pure nitrogen gas (N_2), which than escapes to the atmosphere. The reaction mechanism istriggered by a newly discovered bacterium, called *Brocadia anammoxidans*. Thisappears to be a compartmentalized bacterium; within the cell membrane two compartments can be found which are also surrounded by a membrane, a very rare phenomenon. Intermediate products of the reaction included hydroxylamine, and toxic hydrazine compounds. The bacterial membranes were found to consists of badly permeable membranes, which are thought to function as a barrier for hydrazines produced within the cell. This discovery has major consequences, as it alters the entire contribution of oceans to the nitrogen balance.

PHOSPHORUS CYCLE

The phosphorus cycle involves the uptake of phosphorus by organisms. Phosphorus in the environment is mainly found in rocks, and natural weathering processes can make it available to biological systems. After decomposition of biological waste, it can accumulate in large amounts in soils and sediments. Phosphorus is used by humans as a fertilizer in farmlands and in detergents. Overuse of phosphorus can lead to eutrophication.

PHOSPHORUS CYCLE

Phosphorus is an essential nutrient for plants and animals in the form of ions PO_4^{3-} and HPO_4^{2-}. It is a part of DNA-molecules, of molecules that store energy (ATP and ADP) and of fats of cell membranes. Phosphorus is also a building block of certain parts of the human and animal body, such as the bones and teeth.

Phosphorus can be found on earth in water, soil and sediments. Unlike the compounds of other matter cycles phosphorus cannot be found in air in the gaseous state. This is because phosphorus is usually liquid at normal temperatures and pressures. It is mainly cycling through water, soil and sediments. In the atmosphere phosphorus can mainly be found as very small dust particles. Phosphorus moves slowly from deposits on land and in sediments, to living organisms, and than much more slowly back into the soil and water sediment. The phosphorus cycle is the slowest one of the matter cycles that are described here.

Phosphorus is most commonly found in rock formations and ocean sediments as phosphate salts. Phosphate salts that are released from rocks through weathering usually dissolve in soil water and will be absorbed by plants. Because the quantities of phosphorus in soil are generally small, it is often the limiting factor for plant growth. That is why humans often apply phosphate fertilizers on farmland. Phosphates are also limiting factors for plant-growth in marine ecosystems, because they are not very water-soluble. Animals absorb phosphates by eating plants or plant-eating animals.

Phosphorus cycles through plants and animals much faster than it does through rocks and sediments. When animals and plants die, phosphates will return to the soils or oceans again during decay. After that, phosphorus will end up in sediments or rock formations again, remaining there for millions of years. Eventually, phosphorus is released again through weathering and the cycle starts over.

Humans also cause environmental problems by interfering with the phosphorus cycle. We remove phosphates on certain locations, to apply it on farmland as fertilizers. On these locations the phosphate supply is moving, causing phosphate levels of surrounding land-soils and groundwater to become too high. On farmland where phosphates are applied as fertilizers, plants do not absorb all phosphate. The phosphates end up in water and stream towards lakes and reservoirs, where they cause a phenomenon called eutrophication. Eutrophication means that the water is so rich in nutrients that it causes certain water plants, such as green algae, to grow extensively. As a result the oxygen supply in the water depletes, causing fish and other oxygen-dependent organisms to die and bacteria that are not oxygen dependent (anaerobic) to take over.

A schematic representation of the phosphorus cycle:

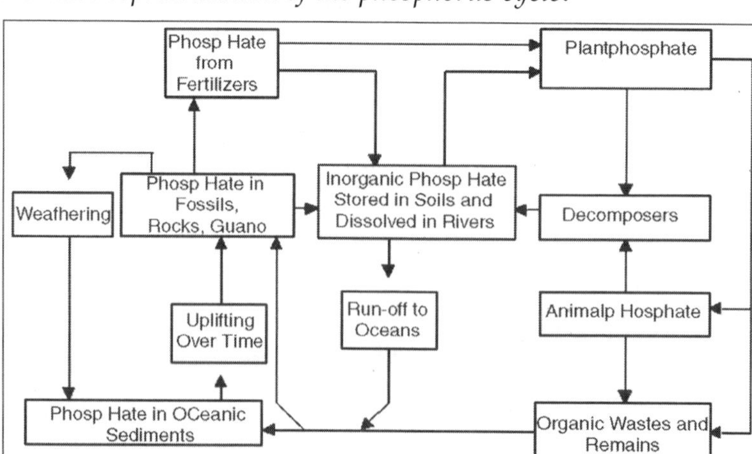

When people practise extensive wood chopping in forest areas, such as the tropical rain forest, phosphate supplies deplete, because most of the phosphate can be found in the ground under the trees in such areas. When the trees are removed the rain washes phosphates away, causing the ground to become unproductive.

Phosphate supplies are restored very slowly, because most phosphate in soils and water comes from weathering rocks and oceanic sediments.

SULPHUR CYCLE

Sulphur is one of the components that make up proteins and vitamins. Proteins consist of amino acids that contain sulphur atoms. Sulphur is important for the functioning of proteins and enzymes in plants, and in animals that depend upon plants for sulphur. Plants absorb sulphur when it is dissolved in water. Animals consume these plants, so that they take up enough sulphur to maintain their health. Most of the earth's sulphur is tied up in rocks and salts or buried deep in the ocean in oceanic sediments. Sulphur can also be found in the atmosphere. It enters the atmosphere through both natural and human sources. Natural recourses can be for instance volcanic eruptions, bacterial processes, evaporation from water, or decaying organisms. When sulphur enters the atmosphere through human activity, this is mainly a consequence of industrial processes where sulphur dioxide (SO_2) and hydrogen sulphide (H_2S) gases are emitted on a wide scale.

When sulphur dioxide enters the atmosphere it will react with oxygen to produce sulphur trioxide gas (SO_3), or with other chemicals in the atmosphere, to produce sulphur salts. Sulphur dioxide may also react with water to produce sulphuric acid (H_2SO_4). Sulphuric acid may also be produced from demethylsulphide, which is emitted to the atmosphere by plankton species. All these particles will settle back onto earth, or react with rain and fall back onto earth as acid deposition. The particles will than be absorbed by plants again and are released back into the atmosphere, so that the sulphur cycle will start over again.

Humans cause environmental problems by influencing the carbon cycle in two ways. Firstly, the removal of forests has caused depletion in plants and trees that absorb carbon dioxide. Humans cause great carbon emissions to the atmosphere, which take place during industrial processes, such as coal and oil combustion. These processes serve the generation of energy. As a result of these effects the carbon dioxide levels in the atmosphere have greatly increased. Between 1870 and 1990 the carbon dioxide emissions to air have increased 25%. The carbon dioxide levels in air are now so high, that the uptake by plants and oceans is not fast enough.

A schematic representation of the sulphur cycle:

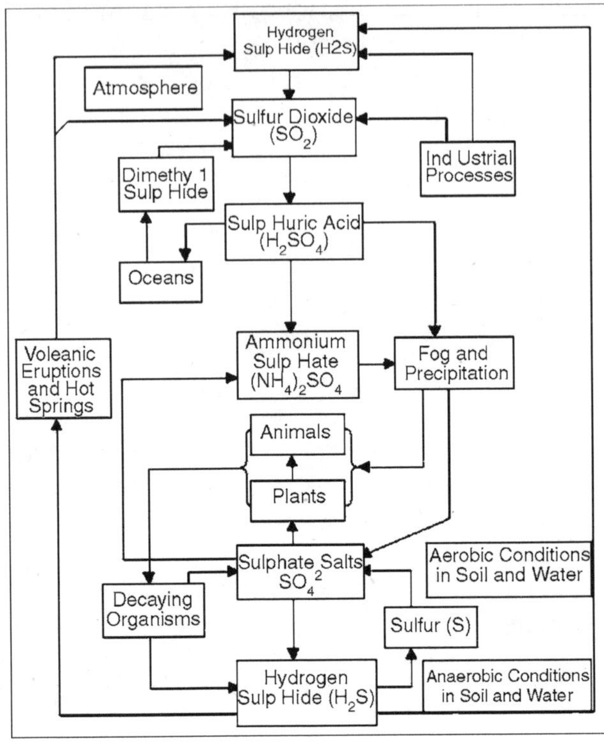

Carbon dioxide is a greenhouse gas. This means that increasing carbon dioxide levels in air support the greenhouse effect. The greenhouse effect is a term that is used for climate change on earth, caused by greenhouse gases in the atmosphere. Too much carbon dioxide in the atmosphere can cause the earth to warm up. This will in time cause all kinds of negative effects, such as melting of ice from the Arctic causing floods on other continents. Secondly, during the industrial revolution humans have used fossil fuels extensively for energy generation, as has been mentioned before. Because fossil fuels consist of dead organic matter it takes a very long time to restore the supplies. Because humans have been burning fossil fuels extensively over the past decades fossil fuel supplies are in danger of being exhausted.

Index

❑❑❑